ADVANTAGE

PRACTICE WORKBOOK

TEACHER'S EDITION

On My Own

Harcourt Brace & Company

Orlando • Atlanta • Austin • Boston • San Francisco • Chicago • Dallas • New York • Toronto • London

http://www.hbschool.com

Printed in the United States of America

ISBN 0-15-307943-6

2 3 4 5 6 7 8 9 10 082 2000 99 98 97

CONTENTS

CHAPTER 23 Experiments with Probability

CHAPTER 24 Measuring Length and Area

CHAPTER 25 Surface Area and Volume

CHAPTER 26 Changing Dimensions

CHAPTER 27 Percent: Spending and Saving

CHAPTER 28 Showing Relationships

Name ____________________

LESSON 1.1

Sets of Numbers

Vocabulary

Write *true* or *false*.

1. Every integer is a rational number. true

2. Every rational number is an integer. false

Classify the numbers as counting numbers (C), whole numbers (W), integers (I), or rational numbers (R). Make a table to show your work. **Check students' tables.**

3. $15, \frac{2}{3}, {}^{-}4$ **C, W, I, R; R; I, R**

4. $0, {}^{-}7.5, \frac{1}{5}$ **W, I, R; R; R**

Tell whether each statement is *true* or *false.* If it is true, draw a Venn diagram to show the relationship. **Check students' diagrams.**

5. Most birds fly. true

6. No birds fly. false

7. Only birds fly. false

For Exercises 8–10, use the Venn diagram.

8. Which counting numbers less than 20 are multiples of 2 and also of 3?

6, 12, 18

9. Which counting numbers less than 20 are odd but not multiples of 3?

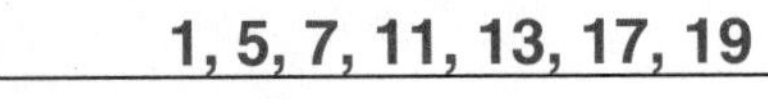

1, 5, 7, 11, 13, 17, 19

Counting Numbers Less Than 20

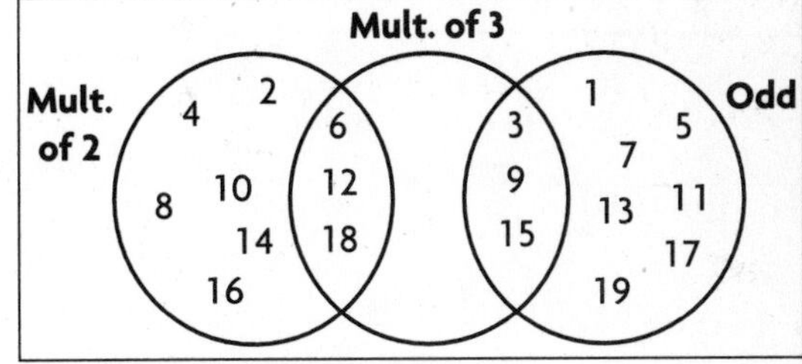

10. How many counting numbers less than 20 are either multiples of 2 or odd, but not both? 19

Mixed Applications

11. The sum of two integers is a whole number. One of the numbers is not a whole number. What can you say about the other number?

It is a whole number.

12. The sum of two integers is a whole number but not a counting number. What is the only whole number that is not a counting number?

0

13. Find the area of Mrs. Maser's 11-ft × 15-ft rectangular garden.

165 ft^2

14. Bobbie is singing 4 songs at a recital. In how many different orders can she sing them?

24 orders

Use with text pages 16–19.

Name ______________________________

LESSON 1.2

Understanding Rational Numbers

Name a rational number for the given point on the number line.

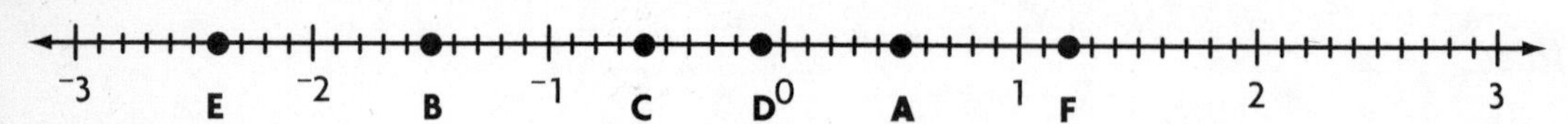

1. point A 0.5
2. point B −1.5
3. point C −0.6
4. point D −0.1
5. point E −2.4
6. point F 1.2

Give at least three other names for each rational number. **Possible answers are given.**

7. 2 — $\frac{4}{2}$, 2.0, $\frac{6}{3}$
8. $\frac{5}{6}$ — $\frac{10}{12}$, $\frac{15}{18}$, $\frac{20}{24}$
9. −3.4 — $-3\frac{2}{5}$, −3.40, $-3\frac{4}{10}$
10. −5 — $\frac{-10}{2}$, −5.0, $\frac{-30}{6}$
11. 4.01 — $4\frac{1}{100}$, 4.010, 4.0100
12. 2.2 — $2\frac{2}{10}$, 2.20, $2\frac{1}{5}$
13. $\frac{-1}{4}$ — −0.25, −0.250, $\frac{-2}{8}$
14. 120 — $\frac{120}{1}$, 120.0, $\frac{240}{2}$

Graph the rational numbers. Use one number line for each exercise. **Check students' number lines.**

15. 0.5, $\frac{1}{4}$, 2.0
16. $\frac{3}{8}$, $\frac{-1}{2}$, −1.7
17. −0.25, 1.75, $\frac{-7}{4}$
18. $\frac{1}{8}$, $\frac{9}{8}$, $\frac{-1}{4}$

Mixed Applications

19. The Philadelphia Phillies won the pennant one year with a winning percentage of 0.599. What is the lowest percentage, to the nearest thousandth, that would have beaten the Phillies that year?

    0.600

20. Three students spend an average of 50 minutes on a homework assignment. Two students spend 42 and 51 minutes. How long does the third student spend on the assignment?

    57 minutes

21. In order to work properly, a piston's diameter must measure between 1.901 in. and 1.904 in. What are three diameter measurements that will work properly?
    Possible answers:

    1.902 in., 1.9025 in., 1.9031 in.

22. The sides of a triangular yard measure 41.3 m, 52.8 m, and 31.5 m. Find the perimeter of the yard.

    125.6 m

Use with text pages 20–21.

Name ________________________________

Parts as Percents

Write percents for the area covered by the small unshaded parts and for the area covered by the small shaded parts.

1. 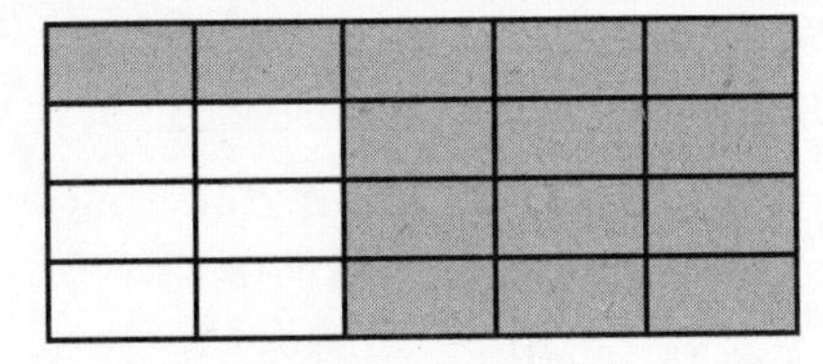

u: 30%; s: 70%

2. 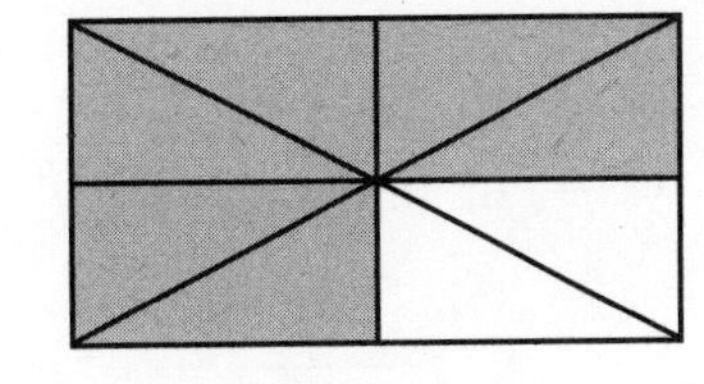

u: 25%; s: 75%

Suppose a 3-in. × 4-in. rectangular piece of paper is folded and then unfolded as shown.

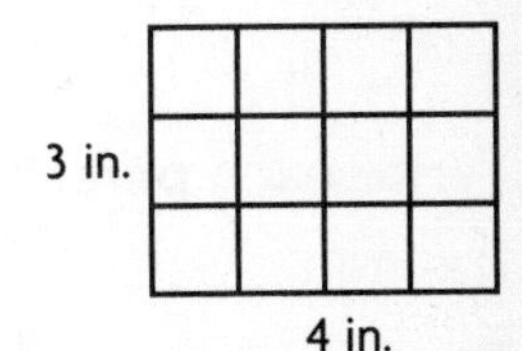

3. How many 1-in. × 1-in. squares are formed? 12 squares

4. How many 1-in. × 3-in. rectangles are formed? 1-in. × 4-in. rectangles?

10 rectangles; 3 rectangles

5. What other size rectangle can be formed? What percent of the whole does its area represent?

1 in. × 2 in.; $\frac{2}{12} = 16.\overline{6}\%$

6. If exactly 3 of the 1-in. × 1-in. squares are colored green, what percent of the paper is green? 25%

Mixed Applications

7. Each piece of a large rectangular cake represents 4% of the whole. Into how many pieces has the cake been cut?

25 pieces

8. Jason gets 19 out of 20 problems correct on a math test. What is his grade on the test expressed as a percent?

95%

9. To raise money for a trip to Washington, D.C., the seventh-grade class does odd jobs. They have earned $650 out of the $2,000 that they need. If they have six weeks left, how much must they earn per week?

$225

10. The flag of Belgium is divided equally into three parts. One part is black, one is yellow, and one is red. What percent of the Belgian flag is red?

$33\frac{1}{3}\%$

Name ______________________

LESSON 1.4

Connecting Fractions, Decimals, and Percents

Write as a percent.

1. $\frac{3}{20}$ **15%**
2. $\frac{3}{4}$ **75%**
3. 0.57 **57%**
4. $\frac{3}{1}$ **300%**
5. 1 out of 200 **0.5%**
6. 0.025 **2.5%**
7. 7 of 8 **87.5%**
8. 97 of 200 **48.5%**
9. $\frac{6}{2}$ **300%**
10. $\frac{1}{8}$ **12.5%**
11. .005 **0.5%**
12. 6 of 16 **37.5%**

Write each percent as a fraction in simplest form.

13. 9% $\frac{9}{100}$
14. 85% $\frac{17}{20}$
15. 40% $\frac{2}{5}$
16. 32% $\frac{8}{25}$
17. 20% $\frac{1}{5}$
18. 98% $\frac{49}{50}$
19. 18% $\frac{9}{50}$
20. 200% **2**

Write each percent as a decimal.

21. 26% **0.26**
22. 90% **0.9**
23. 45.5% **0.455**
24. 3.0% **0.03**
25. 541% **5.41**
26. 8.0% **0.08**
27. 100% **1.00**
28. 12.5% **0.125**

Mixed Applications

29. Jules bought $\frac{1}{50}$ of all the pencils in a bookstore. What percent of the pencils did he purchase?

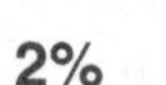

2%

30. Partway through the season, the Orioles have won 31 out of 50 games. What percent of their games have they won?

62%

31. Shawna and her family go to the movies. Adult's tickets cost \$6.50 and children's tickets cost \$4.50. If they spend a total of \$26.50, how many adults and how many children go to the movies?

2 adults, 3 children

32. Derrick scores 17 correct out of 25 problems on a spelling test. What is his score expressed as a fraction, a decimal, and a percent?

$\frac{17}{25}$, **0.68, 68%**

Use with text pages 26–29.

Name ______________________

Making Circle Graphs

Vocabulary

Complete.

1. A(n) circle graph can be used to compare parts with the whole.

Complete the table to find the central angle measures for a circle graph of the survey results.

	Favorite Subject	Percent	Angle Measure
2.	Language Arts	55%	$0.55 \times 360° =$ 198°
3.	Science	20%	0.20 × 360° = 72°
4.	Math	25%	0.25 × 360° = 90°

For Exercises 5–6, use the following percents for the budget of the town of Fairview: schools, 45%; sanitation, 25%; roads, 20%; other, 10%.

5. Calculate the measure of each central angle for a circle graph. Display your results in a table. Check students' tables.

   Measures: 162°; 90°; 72°; 36°

6. Draw a circle graph to represent Fairview's budget. Use a protractor to draw the central angles. Write the categories and the percents on your graph. Give your graph a title.
   Check students' graphs.

Mixed Applications

For Problems 7–8, use the information and your circle graph from Exercises 5–6.

7. If Fairview's budget is $4,500,000, about how much does the town spend for schools?

   about $2,000,000

8. What is the sum of the percents for Fairview's budget? the sum of the degrees of the central angles?

   100%; 360°

9. Rogers Hornsby's batting average as a decimal was 0.383 in 1923 and 0.424 in 1924. Find the difference in batting averages.

   0.041

10. Find the median for the following test scores: 64, 86, 75, 91, 78, 85, 71, 78, 50, 99, 77, 83, 90.

    78

Name ________________________________

LESSON 2.1

Using Exponents

Complete the powers to show two ways to represent the number.

1. $2^? = 512$
   $8^? = 512$
   9; 3

2. $4^? = 256$
   $16^? = 256$
   4; 2

3. $3^? = 6{,}561$
   $9^? = 6{,}561$
   8; 4

4. $5^? = 625$
   $25^? = 625$
   4; 2

Show three ways to represent the number using powers. **Possible answers are given.**

5. 81 — $81^1; 9^2; 3^4$
6. 16 — $16^1; 2^4; 4^2$
7. 160,000 — $160{,}000^1; 400^2; 20^4$
8. 117,649 — $117{,}649^1; 343^2; 7^6$

Find the value. You may want to use a calculator.

9. 2^6 64
10. 5^4 625
11. 25^2 625
12. 21^4 194,481
13. 10^4 10,000
14. 4^0 1
15. 6^5 7,776
16. 1^{12} 1

Write in scientific notation.

17. 5,000 5×10^3
18. 112,000,000 1.12×10^8
19. 23,100 2.31×10^4
20. 3,100,000 3.1×10^6

Write in standard form.

21. 5.7×10^6 5,700,000
22. 1.23×10^4 12,300
23. 9.01×10^{10} 90,100,000,000
24. 8.3×10^5 830,000

Mixed Applications

25. Astronomers recently studied a galaxy that is 11 billion light years from the earth. Expressed in standard form, eleven billion is 11,000,000,000. Write 11 billion in scientific notation.

    1.1×10^{10}

26. Adam is making a circle graph to show his expenses. He spends 30% of his money on books and school supplies. What will be the measure of the central angle for the part of the graph that shows books and school supplies?

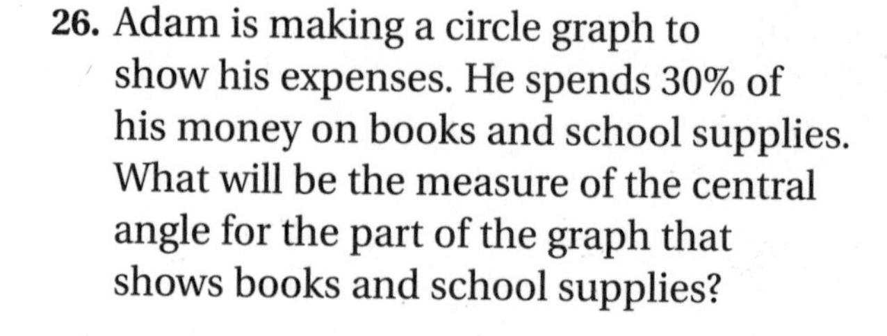

    108°

27. A light year is equal to 5.87×10^{12} mi. Express the number of miles in standard form.

    5,870,000,000,000 mi

28. In scientific notation, how many miles are there in 10 light years?

    5.87×10^{13} mi

Use with text pages 36–39.

Name ______________________

LESSON 2.2

Exploring Decimal and Binary Numbers

Vocabulary

Complete.

1. What is the name of the number system that uses only two digits, 0 and 1?

the binary number system

Use the expanded form to write each binary number as a decimal number.

2. 10_{two} 2
3. 1_{two} 1
4. 1111_{two} 15
5. 1110_{two} 14
6. 1101_{two} 13
7. 10001_{two} 17
8. 10000_{two} 16
9. 10010_{two} 18
10. 11001_{two} 25
11. 11010_{two} 26
12. 1001_{two} 9
13. 11110_{two} 30
14. 1010_{two} 10
15. 11_{two} 3
16. 111_{two} 7
17. 1011_{two} 11

Mixed Applications

18. Use expanded form to find the decimal number for 1101001_{two}.

$(1\times2^6) + (1\times2^5) + (0\times2^4) + (1\times2^3) + (0\times2^2) + (0\times2^1) + (1\times2^0) = (1\times64) + (1\times32) + (0\times16) + (1\times8) + (0\times4) + (0\times2) + (1\times1) = 64 + 32 + 0 + 8 + 0 + 0 + 1 = 105$

19. Using binary code, a computer multiplies 100101_{two} by 11_{two} and finds the answer is 1101111_{two}. What are the decimal numbers? Does the computer find the correct answer?

37; 3; 111; yes

20. An L-shaped boardwalk measures 120 ft in one direction and 600 ft in the other. The width of the boardwalk is 60 ft. What is the area of the boardwalk in square feet? Draw a diagram to help you solve.

39,600 ft^2

21. A baseball team won 57 out of 90 games. Write the winning ratio as a fraction. Then write the fraction as a percent, rounded to one decimal place.

$\frac{57}{90}$; 63.3%

Name ______________________

Modeling Squares and Square Roots

Vocabulary

Complete.

1. A perfect square, also called a(n) ___square number___, is a number that is the square of an integer.
2. A number that can be represented by a geometric figure is called a(n) ___figurate number___.

Tell how many counters you need to make a square array with the given number of counters on one side.

3. 5 ___25___
4. 13 ___169___
5. 30 ___900___
6. 14 ___196___

Suppose a square array is made up of 625 pennies.

7. How many pennies are on each side? on all the sides combined? not on a side? ___25; 96; 529___

Find the square.

8. 16^2 ___256___
9. 40^2 ___1,600___
10. 50^2 ___2,500___
11. 6^2 ___36___

Find the square root.

12. $\sqrt{196}$ ___14___
13. $\sqrt{2{,}500}$ ___50___
14. $\sqrt{2.25}$ ___1.5___
15. $\sqrt{0.81}$ ___0.9___

Locate each square root between two integers.

16. $\sqrt{20}$ ___4 and 5___
17. $\sqrt{91}$ ___9 and 10___
18. $\sqrt{45}$ ___6 and 7___

Mixed Applications

19. The square of a given number is the same as five times the number. What is the number?

___5___

20. The average of Emily's two test scores is 72.5. The scores on both tests are perfect squares. What are the scores?

___81 and 64___

21. A circle graph is titled "Budget." A section with a central angle measuring 43.2° is labeled "Advertising." What percent of the total budget is spent for advertising?

___12%___

22. The area of the earth is approximately 197,000,000 mi^2. Of this, approximately 30% is land. What is the land area of the earth, expressed in scientific notation?

___5.91×10^7 mi^2___

Use with text pages 43–45.

Name ______________________________

Problem-Solving Strategy

Using Guess and Check to Find Square Roots

1. A square of canvas for a tent has an area of 310 ft^2. What is the length of a side, to the nearest tenth of a foot?

   17.6 ft

2. An acre equals 43,560 ft^2. What is the length of a side of a square acre, to the nearest foot?

   209 ft

3. Fred is painting the floor of a square shed. He needs exactly one gallon of paint to cover the 200 ft^2 of floor. What is the length of a side of the shed, to the nearest foot?

   14 ft

4. Hue's basement has a square room with square tiles on the floor. The area of each tile is 2.25 ft^2. The basement floor contains 64 tiles. What is the length of each side of the room?

   12 ft

Mixed Applications

Solve.

CHOOSE A STRATEGY

- Guess and Check
- Solve a Simpler Problem
- Use a Formula
- Write an Equation
- Draw a Diagram

Choices of strategies will vary.

5. The sides of a square are each increased by 20%. The area of the new square is what percent of the area of the original square?

   144%

6. The square of Randolph's age is 117 more than his sister's age. His sister's age is a perfect square. How old is Randolph?

   11 years old

7. The average of four test scores is 81.5. Three of the grades are 80, 83, and 91. What is the fourth grade?

   72

8. Tim is 5 years younger than Juan. Juan is 3 times as old as Sean. Tim is 16 years old. How old is Sean?

   7 years old

9. The formula $d = 0.05 \times s^2 + s$ gives the distance in feet, d, a car travels after the driver puts on the brakes. The variable s represents the speed of the car in miles per hour. If a car is going 50 mi per hour when the brakes are put on, how far will the car continue to travel before it stops?

   175 ft

10. The area of a square plot of land is 3,600 ft^2. Find the area of another square plot in which each side is 1.5 times longer than a side of the first plot.

    8,100 ft^2

Name ______________________________

LESSON 2.5

Repeated Calculations

Vocabulary

Complete.

1. A step in the process of repeating something over and over again is called a(n) ___iteration___.

2. Steps of an iterating process can be shown in a(n) ___iteration diagram___.

For Exercises 3–4, use the iteration process shown. Write the results of the first five iterations.

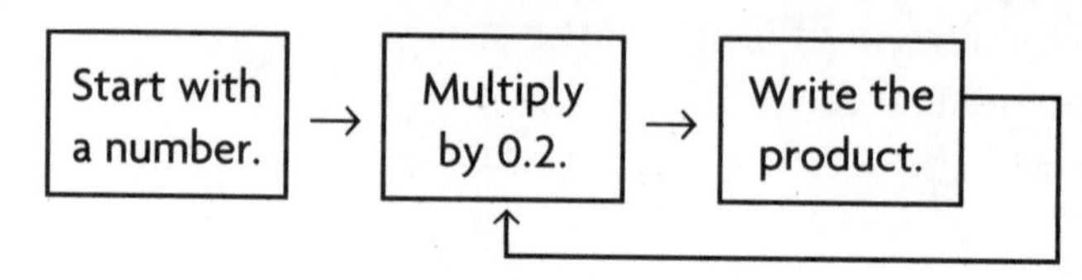

3. Start with 3,600.

___720, 144, 28.8, 5.76, 1.152___

4. Start with 400.

___80, 16, 3.2, 0.64, 0.128___

For Exercises 5–7, use the iteration process shown above.

5. Start with 40. What is the first stage with a result less than 1? ___stage 3___

6. Start with 2,000. At what stage is the number nearest to 10? ___stage 3___

7. If the number at stage 4 is 2, what is the starting number? ___1,250___

Mixed Applications

8. A square is drawn 1 in. on each side. A second square is drawn inside the first by connecting the midpoints of the sides of the square. This is repeated until four squares are constructed inside the original one. If the drawing continues, do you think this is an iteration process? Why or why not?

___Yes; the same step is repeated over and over again.___

9. You invest \$100 and earn 10% interest each year. If you add that 10% to the original principal invested, the new principal at the end of one year will be $\$100 \times 1.1 = \$110$. At the end of the second year, it will be $\$110 \times 1.1 = \$121$. Using this method, about how long will it take \$100 to grow to \$200?

___about 7 years___

10. A building is being constructed on a square piece of land with an area of 200,000 ft^2. Find the length of one side of the plot, rounded to the nearest foot.

___447 ft___

11. The state of Alaska has an area of 656,424 mi^2. If Alaska were a square, what would be the length of a side, rounded to the nearest mile?

___810 mi___

Use with text pages 48–50.

Name ______

LESSON 3.1

Estimating Sums and Differences

During a science experiment, the times it takes different objects to fall from a building to the ground are measured. Estimate the sum of the times of each object by clustering.

1. 15.1 sec, 14.8 sec, 14.6 sec, 15.2 sec, 15.3 sec — **75 sec**

2. 73.4 sec, 71.2 sec, 69.1 sec, 69.7 sec, 72 sec — **350 sec**

3. 50.1 sec, 51.5 sec, 48.9 sec, 49.7 sec, 52.2 sec — **250 sec**

Choose rounding or clustering to estimate each sum. Name the method you used. **Possible estimates and choices of methods are given.**

4. 40.24 in. + 40.74 in. + 39.9 in. + 40.14 in.

   160 in., clustering

5. 82.4 + 67.1 + 102.6 + 32.18

   280, rounding

6. $8.89 + $4.21 + $16.67 + $10.32

   $40.00, rounding

7. 9.3 + 8.9 + 9.2 + 9 + 9.5

   45, clustering

Each group of numbers gives the bowling averages for four members of a bowling team. Find a range for an estimate of the total points scored by these players in a game.

8. Strikers: 100.8, 70.1, 93.6, 112.2

   375 to 380 points

9. Jets: 130.1, 114.7, 84.6, 102.3

   430 to 484 points

Estimate the difference by rounding. **Possible estimates are given.**

10. 5.2 − 3.48 — **2**

11. 47,843 − 19,231 — **29,000**

12. 184.13 − 69.89 — **110**

13. 3,432 − 761.9 — **2,600**

Mixed Applications

14. Alex has a budget of $75.00 for sporting goods. He wants to buy items that cost $9.35, $18.80, $11.09, $15.79, $10.75, $4.60, and $2.56. Estimate to see if he can buy all of the items.

    Yes. The total cost is about $74.00.

15. Marcy is 3 years younger than Brooke, who is four times as old as Stuart. Marcy is 29. How old is Stuart?

    8 years old

Use with text pages 60–63.

Name ______________________________

Multiplying Whole Numbers and Decimals

Use decimal squares to find the product. **Check students' work.**

1. 3×0.4

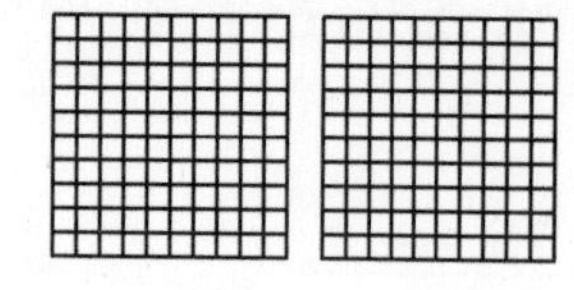

1.2

2. 0.8×0.3

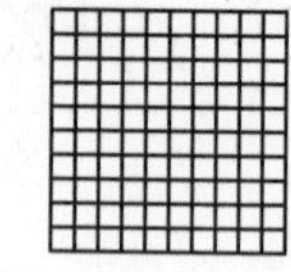

0.24

Place the decimal point in the product. Write zeros as needed.

3. $\begin{array}{r} 32.6 \\ \times\ 7 \\ \hline 2282 \end{array}$ **228.2**

4. $\begin{array}{r} 45.3 \\ \times\ 0.24 \\ \hline 10872 \end{array}$ **10.872**

5. $\begin{array}{r} 0.009 \\ \times\ 0.04 \\ \hline 36 \end{array}$ **0.000**36

6. $\begin{array}{r} 0.007 \\ \times\ 8 \\ \hline 56 \end{array}$ **0.0**56

Find the product.

7. 19.8×3 **59.4**

8. 4.9×2 **9.8**

9. 5×8.6 **43**

10. 4.28×6 **25.68**

11. 3.74×0.8 **2.992**

12. 8.29×0.9 **7.461**

13. 35×0.06 **2.1**

14. 73.85×1.03 **76.0655**

15. 1.6×0.82 **1.312**

16. 67.5×5.3 **357.75**

17. 5.83×0.93 **5.4219**

18. 8.34×4.7 **39.198**

19. 0.0004×6.6 **0.00264**

20. 1.62×0.031 **0.05022**

21. 38.45×4.3 **165.335**

22. 96.2×0.89 **85.618**

Mixed Applications

23. Admission to the Shedd Aquarium costs $5.50 for adults and $2.25 for students. Mrs. Johnson takes her class of 24 students to the aquarium. How much does the class spend on tickets?

$59.50

24. The seventh graders at Crystal Springs School have a bake sale. They sell 0.875 of the baked goods. What fraction of the baked goods do they sell? What percent?

$\frac{7}{8}$; 87.5%

25. To raise money for his scout troop, Mike sold 16 cans of popcorn for $6.50 each and 18 calendars for $3.75 each. What were Mike's total sales?

$171.50

26. At a gymnastics meet, Missy must earn a total of 34 points to advance to a higher level of competition. Four judges give her the following scores: 9.1, 8.9, 8.7, and 9.3. Does Missy advance? What is Missy's estimated total?

yes; 36 points

Use with text pages 64–66.

Name ____________________

Dividing Whole Numbers and Decimals

Use decimal squares to find the quotient. **Check students' work.**

1. $1.2 \div 6$

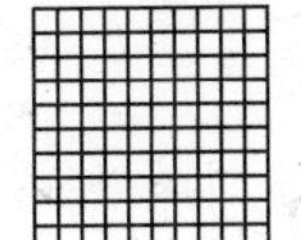

0.2

2. $0.16 \div 2$

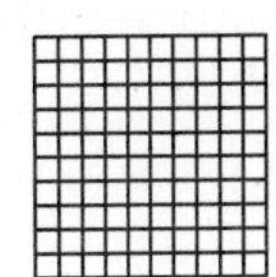

0.08

Find the quotient.

3. $56.48 \div 8$ **7.06**
4. $70.2 \div 9$ **7.8**
5. $49.446 \div 8.2$ **6.03**
6. $28.98 \div 6.3$ **4.6**

A regular polygon and its perimeter are given. Find the length of a side.

7. nonagon, 57.15 ft **6.35 ft**
8. pentagon, 466.5 in. **93.3 in.**
9. decagon, 942.16 in. **94.216 in.**

Find the speed in miles per hour by dividing the distance (d) by the time (t).

10. $d = 33.25$ mi, $t = 2.5$ hr **13.3 mph**
11. $d = 376.2$ mi, $t = 4.5$ hr **83.6 mph**
12. $d = 255.84$ mi, $t = 10.4$ hr **24.6 mph**
13. $d = 208.15$ mi, $t = 5.75$ hr **36.2 mph**

Estimate the quotient by using compatible numbers. **Possible estimates are given.**

14. $44.9 \div 9.32$ **about 5**
15. $12{,}089.4 \div 42.34$ **about 300**
16. $152.6 \div 14.89$ **about 10**
17. $598.3 \div 1.8$ **about 300**

Mixed Applications

18. Sasha must pay off a $6,542.86 credit card balance. She can afford to pay no more than $300.00 per month. Can she pay off the loan in 24 equal monthly payments? Explain.

Yes. Each payment is $272.62.

19. The batting average of a baseball player is the number of hits made divided by the number of times at bat. If Tom made 91 hits while at bat 285 times, what is his batting average to the nearest thousandth?

0.319

20. The area of a square pigpen is 1,000 ft^2. Between what two integers is the measure of a side?

31 ft and 32 ft

21. Barbara bought 16 packages of dried fruit to bake cookies for the school bake sale. Each package cost $2.49. What was the cost of the dried fruit?

$39.84

Name ________________________

Order of Operations

1. Follow this sequence of operations to find the value.
square → multiply → divide → subtract → add $3 \times 3 - 3^2 + 3 \div 3$ **1**

2. Follow this sequence of operations to find the value.
cube → divide → multiply → add → subtract $4 \div 4 + 4 \times 4^3 - 4$ **253**

Give the correct order of operations for finding the value, and then find the value.

3. $6 \div 2 + 36 \times 3$
÷, ×, +; 111

4. $40 \times (6 - 2)$
−, ×; 160

5. $40 \times 6 - 2$
×, −; 238

6. $5^2 + 16 \div 4 - 3 \times 2$
exponent, ÷, ×, +, −; 23

7. $(6^2 - 2) \div (36 - 34)$
exponent, −, −, ÷; 17

8. $(3^3 + 9 + 12) - 6 \times 2$
exponent, +, +, ×, −; 36

9. $4 \div (4 + 4) \times 4^3 - 4$
+, exponent, ÷, ×, −; 28

10. $4 \div 4 + (4 \times 4)^3 - 4$
×, exponent, ÷, +, −; 4,093

11. $4 \div 4 + 4 \times (4^3 - 4)$
exponent, −, ÷, ×, +; 241

Find the value by using a calculator.

12. $14.3 \times 2.1 - 6.2 \times 1.5$
20.73

13. $(4.8 - 1.2) \times (8 - 6)^2$
14.4

14. $24 - 5 + 4^2 \times 3$
67

Mixed Applications

15. Mrs. Hess bought 3 lb of coffee at $4.99 per pound and 5 lb of coffee at $5.99 per pound. How much did Mrs. Hess spend on coffee?
$44.92

16. Carlos has dimes and quarters in his coin holder. The 12 coins have a value of $2.55. How many of each coin does he have?
3 dimes and 9 quarters

17. Bill's paycheck from Jewel Food Store is $83.70 for 13.5 hr. How much does Bill earn per hour?
$6.20 per hour

18. Mr. Wynn calculates the number of calories in every meal. He eats 2 hot dogs and 25 French fries for lunch. Each hot dog has 250 calories, and each French fry has 20 calories. How many calories does he eat for lunch?
1,000 calories

Use with text pages 71–73.

Name ______________________________

LESSON 4.1

Adding and Subtracting Fractions

Vocabulary

Complete.

1. The greatest factor that two or more numbers have in common is called the **greatest common factor, or GCF**.

2. The smallest common multiple of two or more denominators is called the **least common denominator, or LCD**.

Add. Write the answer in simplest form.

3. $\frac{2}{5} + \frac{1}{8}$ **$\frac{21}{40}$**

4. $\frac{1}{2} + \frac{1}{6}$ **$\frac{2}{3}$**

5. $\frac{3}{7} + \frac{2}{3}$ **$\frac{23}{21}$, or $1\frac{2}{21}$**

6. $\frac{5}{6} + \frac{1}{10}$ **$\frac{14}{15}$**

7. $\frac{5}{12} + \frac{1}{4}$ **$\frac{2}{3}$**

8. $\frac{3}{5} + \frac{1}{2} + \frac{3}{10}$ **$\frac{7}{5}$, or $1\frac{2}{5}$**

9. $\frac{4}{15} + \frac{1}{3} + \frac{3}{10}$ **$\frac{9}{10}$**

10. $\frac{3}{8} + \frac{1}{6} + \frac{5}{12} + \frac{2}{3}$ **$\frac{13}{8}$, or $1\frac{5}{8}$**

Subtract. Write the answer in simplest form.

11. $\frac{17}{30} - \frac{7}{30}$ **$\frac{1}{3}$**

12. $\frac{3}{4} - \frac{1}{5}$ **$\frac{11}{20}$**

13. $\frac{1}{2} - \frac{3}{16}$ **$\frac{5}{16}$**

14. $\frac{9}{10} - \frac{2}{5}$ **$\frac{1}{2}$**

15. $\frac{5}{6} - \frac{4}{7}$ **$\frac{11}{42}$**

16. $\frac{19}{24} - \frac{5}{8}$ **$\frac{1}{6}$**

17. $\frac{11}{12} - \frac{5}{9}$ **$\frac{13}{36}$**

18. $\frac{33}{40} - \frac{5}{8}$ **$\frac{1}{5}$**

19. $\frac{7}{15} - \frac{5}{12}$ **$\frac{1}{20}$**

20. $\frac{2}{3} - \frac{5}{12}$ **$\frac{1}{4}$**

21. $\frac{1}{2} - \frac{1}{5}$ **$\frac{3}{10}$**

22. $\frac{2}{3} - \frac{1}{6}$ **$\frac{1}{2}$**

Mixed Applications

23. Let 1 equal Gilbert's total paycheck. Each week he saves $\frac{1}{10}$ of the paycheck and spends $\frac{3}{5}$ of it on a car payment and insurance. What fraction of Gilbert's paycheck is left?

$\frac{3}{10}$

24. Let 1 equal the total amount of money the seventh-grade class earns at a car wash. If $\frac{1}{3}$ of the earnings pays for supplies and $\frac{1}{5}$ pays for advertising, what fraction of the total is profit?

$\frac{7}{15}$

25. Luz buys nuts that cost \$2.98 per pound. She has \$10.43 in her wallet. How many pounds of nuts can she buy?

3.5 lb

26. William purchases $2\frac{1}{2}$ lb of apples. If the apples cost \$0.76 per pound, how much does William spend on apples?

\$1.90

Name ______________________

LESSON 4.2

Adding and Subtracting Mixed Numbers

Add. Write the answer in simplest form.

1. $3\frac{1}{12} + 4\frac{3}{4}$ $7\frac{5}{6}$

2. $5\frac{1}{8} + 2\frac{9}{10}$ $8\frac{1}{40}$

3. $1\frac{7}{12} + 6\frac{2}{3}$ $8\frac{1}{4}$

4. $5\frac{3}{5} + 7\frac{7}{10}$ $13\frac{3}{10}$

5. $1\frac{4}{5} + 2\frac{1}{3}$ $4\frac{2}{15}$

6. $2\frac{5}{9} + 4\frac{5}{6}$ $7\frac{7}{18}$

7. $3\frac{1}{6} + 4\frac{3}{8}$ $7\frac{13}{24}$

8. $2\frac{11}{12} + 1\frac{3}{4}$ $4\frac{2}{3}$

9. $1\frac{5}{6} + 2\frac{2}{9}$ $4\frac{1}{18}$

10. $3\frac{1}{4} + 4\frac{7}{8}$ $8\frac{1}{8}$

11. $7\frac{3}{8} + 1\frac{5}{6}$ $9\frac{5}{24}$

12. $3\frac{2}{3} + 1\frac{3}{4}$ $5\frac{5}{12}$

Subtract. Write the answer in simplest form.

13. $11 - 5\frac{7}{8}$ $5\frac{1}{8}$

14. $3\frac{1}{6} - 1\frac{5}{8}$ $1\frac{13}{24}$

15. $5\frac{1}{3} - 2\frac{1}{12}$ $3\frac{1}{4}$

16. $8\frac{2}{3} - 3\frac{4}{5}$ $4\frac{13}{15}$

17. $2\frac{1}{4} - 1\frac{3}{8}$ $\frac{7}{8}$

18. $9\frac{7}{10} - 3\frac{3}{4}$ $5\frac{19}{20}$

19. $6\frac{1}{5} - 2\frac{7}{10}$ $3\frac{1}{2}$

20. $12\frac{6}{7} - 10\frac{1}{2}$ $2\frac{5}{14}$

21. $15 - 5\frac{6}{7}$ $9\frac{1}{7}$

22. $8\frac{1}{6} - 7\frac{1}{3}$ $\frac{5}{6}$

23. $12\frac{1}{4} - 9\frac{1}{2}$ $2\frac{3}{4}$

24. $10\frac{3}{8} - 2\frac{1}{4}$ $8\frac{1}{8}$

25. $4\frac{3}{8} - 2\frac{5}{6}$ $1\frac{13}{24}$

26. $11\frac{1}{8} - 10\frac{3}{4}$ $\frac{3}{8}$

27. $7\frac{2}{3} - 1\frac{5}{6}$ $5\frac{5}{6}$

28. $3\frac{3}{8} - 1\frac{1}{4}$ $2\frac{1}{8}$

Mixed Applications

29. The distance from Todd's home to Marlboro Middle School is $2\frac{2}{3}$ mi. The distance from school to the soccer field is $1\frac{1}{5}$ mi. How far does Todd travel from home to school to the soccer field and back to school?

$5\frac{1}{15}$ mi

30. Anita counts the number of maple trees her family taps to make syrup. From different tree lots she counts 34, 15, 48, 32, and 12 trees. A pail collects 5 gal of sap from each tree. Estimate the number of gallons of sap Anita's family will collect.

about 700 gal

31. When an $8\frac{1}{4}$-lb ham is trimmed of fat, it weighs $7\frac{3}{8}$ lb. How much does the fat weigh?

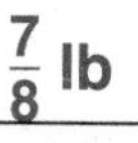

$\frac{7}{8}$ lb

32. The Trojans have 60 yd to go before scoring a touchdown. They gain 19 yd and then lose 7 yd. How far do the Trojans now have to go before scoring a touchdown?

48 yd

Use with text pages 80–81.

Name ____________________

LESSON 4.3

Estimating Sums and Differences

Estimate the sum or difference. **Possible estimates are given.**

1. $\frac{3}{8} + \frac{1}{4}$ $\frac{1}{2}$
2. $\frac{6}{7} + \frac{1}{5}$ 1
3. $\frac{9}{11} - \frac{1}{9}$ 1
4. $1\frac{1}{8} - \frac{7}{8}$ 0
5. $3\frac{7}{12} + 1\frac{5}{8}$ 5
6. $2\frac{1}{8} - 1\frac{7}{16}$ $\frac{1}{2}$
7. $3\frac{2}{3} + 1\frac{1}{10}$ 5
8. $4\frac{2}{5} + 6\frac{1}{12}$ $10\frac{1}{2}$
9. $8\frac{1}{11} - 2\frac{3}{4}$ 5
10. $9\frac{3}{5} - 6\frac{4}{9}$ 3
11. $3\frac{7}{9} + 4\frac{13}{15}$ 9
12. $2\frac{4}{7} + 3\frac{11}{12}$ $6\frac{1}{2}$
13. $11\frac{2}{17} - 4\frac{7}{8}$ 6
14. $9\frac{6}{13} - 1\frac{13}{14}$ $7\frac{1}{2}$
15. $9\frac{1}{20} + 5\frac{2}{35}$ 14
16. $4\frac{5}{9} + 6\frac{6}{11}$ 11
17. $12\frac{1}{6} - 3\frac{3}{4}$ 8
18. $15\frac{7}{12} - 11\frac{5}{9}$ 4
19. $3\frac{11}{12} + 1\frac{1}{8} + 5\frac{5}{12}$ $10\frac{1}{2}$
20. $\frac{3}{8} - \frac{7}{12}$ 0
21. $3\frac{9}{10} + 1\frac{1}{8} + 5\frac{8}{17}$ $10\frac{1}{2}$
22. $14\frac{5}{12} - 6\frac{1}{10}$ $8\frac{1}{2}$
23. $8\frac{13}{27} - 1\frac{11}{12}$ $6\frac{1}{2}$
24. $1\frac{1}{9} + 2\frac{9}{10} + 4\frac{1}{8}$ 8
25. $9\frac{8}{15} - 6\frac{1}{7}$ $3\frac{1}{2}$
26. $1\frac{4}{9} + 3\frac{11}{12} + 2\frac{7}{16}$ 8
27. $16\frac{9}{20} - 8\frac{4}{5}$ $7\frac{1}{2}$
28. $9\frac{1}{16} - 5\frac{13}{24}$ $3\frac{1}{2}$

Mixed Applications

29. Ron buys $1\frac{3}{4}$ lb of pecans, $2\frac{1}{10}$ lb of almonds, and $3\frac{4}{5}$ lb of peanuts. About how many pounds of nuts does Ron buy?

about 8 lb

30. Natasha has $4\frac{7}{8}$ yd of fabric. She needs $1\frac{3}{5}$ yd to make a pillow. About how much extra fabric will Natasha have after making the pillow?

about $3\frac{1}{2}$ yd

31. A circle graph of the Kelly family's budget shows the following expenses: rent, 40%; food, 20%; car, 15%; clothing, 5%; savings, 10%; other, 10%. Calculate the measure of each central angle.

rent, 144°; food, 72°; car, 54°; clothing, 18°; savings, 36°; other, 36°

32. Oranges cost $1.29 per pound at McWhorter's Grocery. Gina paid $6.45 for oranges. How many pounds of oranges did Gina buy?

5 lb

Name ______________________

LESSON 4.4

Multiplying and Dividing Fractions and Mixed Numbers

Vocabulary

Write *true* or *false*.

1. The reciprocal is the number you get when you exchange the numerator and the denominator of a fraction. **true**

Multiply. Write the answer in simplest form.

2. $\frac{2}{7} \times \frac{14}{15}$ **$\frac{4}{15}$**
3. $\frac{3}{8} \times \frac{4}{5}$ **$\frac{3}{10}$**
4. $3\frac{2}{3} \times \frac{9}{22}$ **$\frac{3}{2}$, or $1\frac{1}{2}$**
5. $1\frac{1}{4} \times 8$ **10**

Write the related multiplication problem.

6. $\frac{2}{3} \div \frac{1}{5} = 3\frac{1}{3}$ **$\frac{2}{3} \times \frac{5}{1} = 3\frac{1}{3}$**
7. $\frac{1}{6} \div \frac{2}{3} = \frac{1}{4}$ **$\frac{1}{6} \times \frac{3}{2} = \frac{1}{4}$**
8. $16 \div 4 = 4$ **$16 \times \frac{1}{4} = 4$**
9. $8 \div \frac{2}{5} = 20$ **$8 \times \frac{5}{2} = 20$**

Write the reciprocal of each number.

10. $6\frac{2}{3}$ **$\frac{3}{20}$**
11. 9 **$\frac{1}{9}$**
12. $7\frac{4}{5}$ **$\frac{5}{39}$**
13. $3\frac{3}{4}$ **$\frac{4}{15}$**
14. $\frac{7}{12}$ **$\frac{12}{7}$**

Divide. Write the answer in simplest form.

15. $5 \div \frac{2}{5}$ **$\frac{25}{2}$, or $12\frac{1}{2}$**
16. $\frac{1}{2} \div \frac{1}{8}$ **4**
17. $3\frac{1}{3} \div \frac{9}{10}$ **$\frac{100}{27}$, or $3\frac{19}{27}$**
18. $2\frac{3}{8} \div 1\frac{1}{4}$ **$\frac{19}{10}$, or $1\frac{9}{10}$**
19. $\frac{7}{18} \div \frac{21}{15}$ **$\frac{5}{18}$**
20. $\frac{2}{3} \div \frac{2}{9}$ **3**
21. $\frac{5}{12} \div \frac{1}{3}$ **$\frac{5}{4}$, or $1\frac{1}{4}$**
22. $1\frac{3}{7} \div 1\frac{1}{4}$ **$\frac{8}{7}$, or $1\frac{1}{7}$**

Mixed Applications

23. A bookshelf in Ross's den is 45 in. long. How many books that are $1\frac{1}{2}$ in. thick can fit on the shelf? **30 books**

24. Maria lays 42 pieces of a puzzle end to end. Each piece is $\frac{5}{6}$ in. long. What is the total length of the 42 pieces? **35 in.**

25. Ms. Margoles's weekly salary is $375. Her deductions are $101.25. What percent of Ms. Margoles's salary is deducted? **27%**

26. Mike walks $7\frac{1}{2}$ blocks west, then 2 blocks south, then another $7\frac{1}{2}$ blocks west. How many blocks has Mike walked? **17 blocks**

Use with text pages 86–89.

Name ________________________________

Problem-Solving Strategy

Solve a Simpler Problem

Solve a simpler problem.

1. Thelma, Raja, and Louise travel 14,000 mi during their vacation. Thelma drives $\frac{2}{7}$ of the distance. Louise drives $\frac{1}{2}$ of the remaining distance. How far does Louise drive?

   5,000 mi

2. Sean owns 800 baseball cards. He collected $\frac{2}{5}$ of them before he was ten years old and $\frac{1}{3}$ of the remaining ones before he was twelve years old. How many cards did he collect when he was between the ages of ten and twelve?

   160 cards

3. The Tie Shop has 4,200 ties at the beginning of the month. The owner puts $\frac{1}{6}$ of the ties away for a special sale. She then sells $\frac{4}{7}$ of the remaining ties. How many ties does she sell?

   2,000 ties

4. Raphael earns $750 per month. He saves $\frac{2}{5}$ of his monthly earnings. At this rate, how much will Raphael save in one month? in one year?

   $300; $3,600

Mixed Applications

Solve.

CHOOSE A STRATEGY

- **Find a Pattern**
- **Work Backward**
- **Make a Table**
- **Use a Formula**

Choices of strategies will vary.

5. Today Susan's house is worth 150% of what her family paid for it in 1980. If it is worth $81,000 today, what did they pay for it in 1980?

   $54,000

6. Edwardo went to the beach on May 2, May 3, May 5, May 8, and May 12. If he continues this pattern, when will he visit the beach again?

   May 17

7. In a survey of 100,000 people, a Lin News Agency poll found that $\frac{1}{4}$ of the people surveyed had grey hair. Of these, $\frac{3}{5}$ had curly hair. How many people had curly hair?

   15,000 people

8. Hender's garden is 8 ft long and $5\frac{1}{4}$ ft wide. He wants to cover the garden with netting. How many square feet of netting does Hender need to buy? $A = lw$

   42 ft^2

9. Jane is 9 years older than her sister, who is $\frac{3}{5}$ the age of their brother Ted. If Ted is 15 years old, how old is Jane?

   18 years old

10. Michelle's room is 12 ft long and 14 ft wide. She wants to put a border along the four walls. How many feet of border does she need?

    52 ft

Use with text pages 90–91.

Name ______________________

LESSON 5.1

Adding Integers

Write the addition equation modeled on the number line.

1.

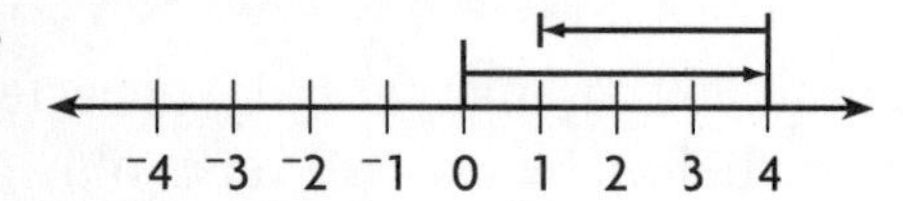

4 + ⁻3 = 1

2.

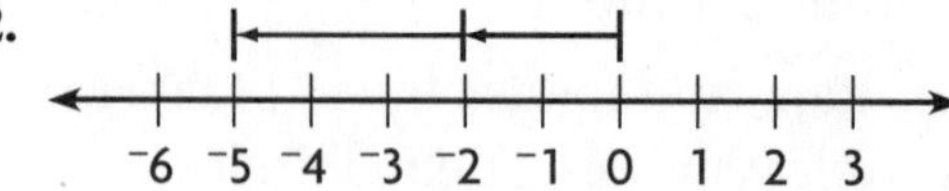

⁻2 + ⁻3 = ⁻5

3.

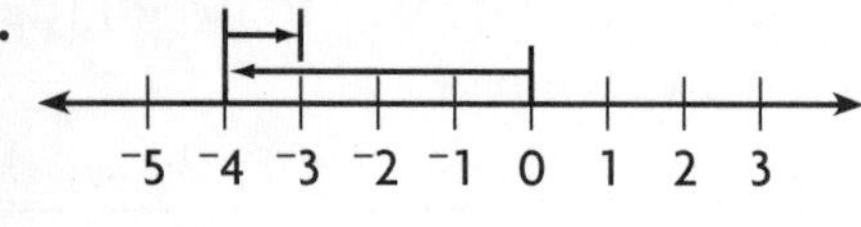

⁻4 + 1 = ⁻3

4.

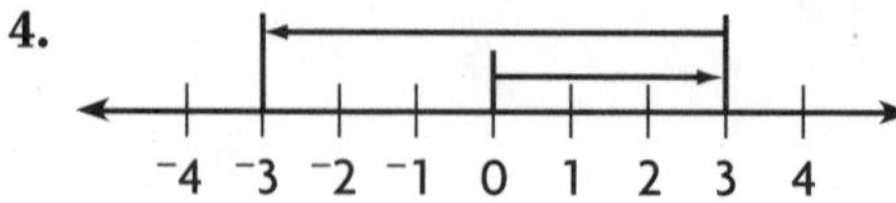

3 + ⁻6 = ⁻3

Draw a number line to find the sum.

5. ⁻8 + ⁻2 ⁻10

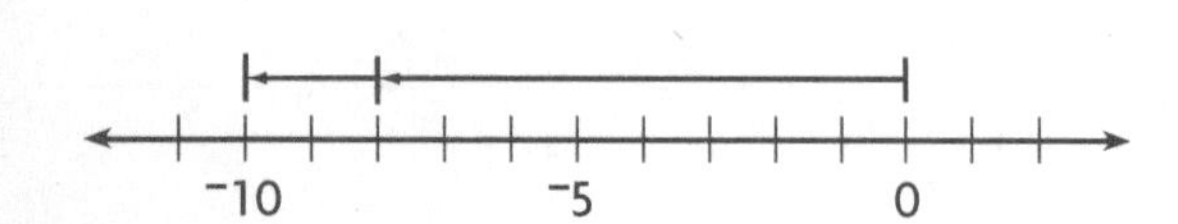

6. 6 + ⁻6 0

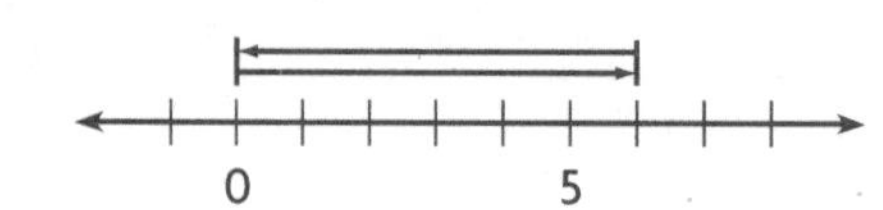

Give the absolute value of each.

7. |9| 9

8. |⁻3| 3

9. |⁻51| 51

10. |27| 27

Find the sum by using the absolute values.

11. 7 + ⁻2 5

12. ⁻5 + 15 10

13. ⁻21 + ⁻13 ⁻34

14. ⁻2 + ⁻15 ⁻17

15. ⁻2 + 13 11

16. ⁻10 + ⁻9 ⁻19

17. 2 + ⁻2 + ⁻2 ⁻2

18. ⁻4 + 1 + ⁻3 ⁻6

Mixed Applications

19. The temperature drops from 21°F to ⁻19°F. What is the temperature change?

⁻40°F or 40°F

20. The central angle of part of a circle graph measures 36°. This represents what percent of the circle?

10%

21. Angel borrowed $141 to buy a stereo. He pays back $14 each week. How much does he still owe after 2 weeks?

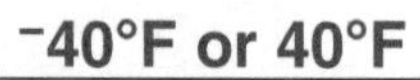

22. A picture with an area of 90 cm² is enlarged so the new picture is 50% larger in area. What is the area of the new picture?

135 cm²

Use with text pages 97–99.

Name ______________________

Subtracting Integers

Write an addition equation and a solution for each of the following.

1. $^-3 - {}^-7 = n$
   $^-3 + 7 = n$;
   $n = 4$

2. $4 - {}^-8 = n$
   $4 + 8 = n$;
   $n = 12$

3. $^-5 - 6 = n$
   $^-5 + {}^-6 = n$;
   $n = {}^-11$

4. $^-9 - {}^-4 = n$
   $^-9 + 4 = n$;
   $n = {}^-5$

5. $5 - {}^-8 = n$
   $5 + 8 = n$;
   $n = 13$

6. $4 - {}^-10 = n$
   $4 + 10 = n$;
   $n = 14$

7. $^-5 - {}^-5 = n$
   $^-5 + 5 = n$;
   $n = 0$

8. $^-3 - 6 = n$
   $^-3 + {}^-6 = n$;
   $n = {}^-9$

Find the difference.

9. $^-13 - 9$ $^-22$
10. $12 - {}^-5$ 17
11. $^-14 - {}^-7$ $^-7$
12. $^-5 - 5$ $^-10$
13. $^-8 - {}^-8$ 0
14. $6 - {}^-7$ 13
15. $^-9 - 14$ $^-23$
16. $2 - 10$ $^-8$
17. $4 - {}^-9$ 13
18. $^-5 - 12$ $^-17$
19. $6 - {}^-3$ 9
20. $^-5 - 5$ $^-10$

Find the difference. Use a calculator.

21. $17 - 87$ $^-70$
22. $^-54 - {}^-37$ $^-17$
23. $75 - {}^-87$ 162
24. $^-63 - {}^-39$ $^-24$
25. $^-101 - 38$ $^-139$
26. $134 - {}^-36$ 170
27. $^-98 - {}^-47$ $^-51$
28. $71 - 199$ $^-128$

Mixed Applications

29. A submarine submerged at a depth of $^-40$ ft dives 57 ft more. What is the new depth of the submarine?

    $^-97$ ft

30. An airplane at 20,000 ft drops 2,500 ft in altitude. What is its new altitude?

    17,500 ft

31. In scientific notation, the speed of light is 1.86×10^5 mi per second. In standard form, how far does light travel in 1 min?

    11,160,000 mi

32. The area of Harvey's Lumber Yard is 30,000 ft^2. The yard is square. Harvey has 640 ft of fence to fence it in. Does he have enough? How much does he need?

    no; 693 ft

Name ______________________

LESSON 5.3

Multiplying and Dividing Integers

1. Show how to find the product $5 \times {}^{-}2$ on a number line.

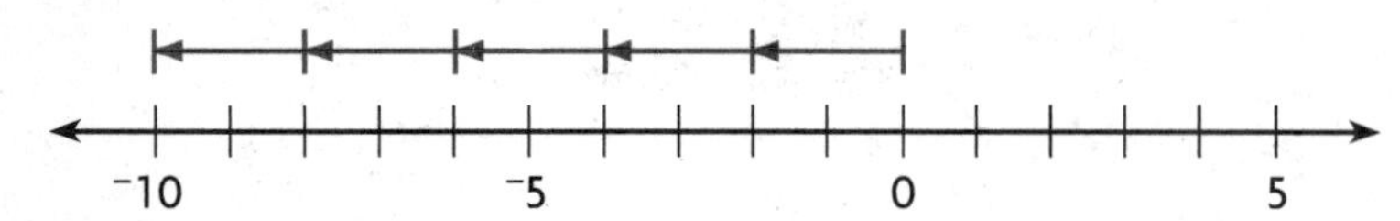

Find the product.

2. $^{-}3 \times 7$ $^{-}21$

3. $6 \times {}^{-}2$ $^{-}12$

4. $5 \times {}^{-}8$ $^{-}40$

5. $^{-}2 \times {}^{-}3$ 6

6. $^{-}3 \times 0$ 0

7. $^{-}5 \times {}^{-}10$ 50

8. $7 \times {}^{-}8$ $^{-}56$

9. $^{-}9 \times 6$ $^{-}54$

10. $^{-}12 \times 9$ $^{-}108$

11. $^{-}10 \times {}^{-}10$ 100

12. $^{-}15 \times {}^{-}5$ 75

13. $8 \times {}^{-}19$ $^{-}152$

14. $^{-}30 \times 6$ $^{-}180$

15. $^{-}1 \times {}^{-}102$ 102

16. $67 \times {}^{-}14$ $^{-}938$

17. $^{-}35 \times {}^{-}13$ 455

18. 74×5 370

19. $^{-}46 \times 23$ $^{-}1{,}058$

20. $^{-}4 \times {}^{-}5 \times 0$ 0

21. $2 \times {}^{-}3 \times {}^{-}8$ 48

22. $^{-}1 \times {}^{-}6 \times {}^{-}9 =$ $^{-}54$

Complete the table.

	Division Problem	Related Multiplication Problem
23.	$^{-}56 \div 8 =$ $^{-}7$	$8 \times$ $^{-}7$ $= {}^{-}56$
24.	$^{-}54 \div {}^{-}9 =$ 6	$^{-}9 \times$ 6 $= {}^{-}54$
25.	$^{-}225 \div {}^{-}15 =$ 15	$^{-}15 \times$ 15 $= {}^{-}225$

Mixed Applications

26. Divers found three treasures in Deep Cove at depths of $^{-}55$ ft, $^{-}48$ ft, and $^{-}71$ ft. Find the average depth at which treasures were found.

$^{-}58$ ft

27. Cody and Jill play four games in which scores can be positive or negative. Cody's scores are $^{-}23$, $^{-}72$, $^{-}49$, and $^{-}52$. What is his average score?

$^{-}49$

28. A large cake for a class picnic is cut into 50 pieces. Each piece of cake is what percent of the whole cake?

2%

29. Jamie spends 1 hr and 20 min doing her homework. Then she spends $\frac{1}{4}$ as much time practicing the violin. How many minutes does she spend practicing the violin?

20 min

Use with text pages 104–106.

Name ____________________

Adding and Subtracting Rational Numbers

Use the rules for adding rational numbers to tell whether the sum is positive or negative. Do not add.

1. $9.5 + {}^-4.7$ positive

2. ${}^-13.1 + {}^-6.5$ negative

3. ${}^-7.6 + 4.4$ negative

4. ${}^-17.3 + {}^-3.3$ negative

Find the sum.

5. ${}^-5.6 + {}^-2.7$ ${}^-8.3$

6. ${}^-8.1 + 3.7$ ${}^-4.4$

7. $1\frac{1}{2} + {}^-\frac{2}{3}$ $\frac{5}{6}$

Find the difference.

8. $23.4 - {}^-11.2$ 34.6

9. ${}^-34.2 - 24.8$ ${}^-59$

10. ${}^-3\frac{1}{4} - {}^-1\frac{1}{2}$ $\frac{^-7}{4}$ or ${}^-1\frac{3}{4}$

Find the sum or difference.

11. $45.2 - {}^-32.8$ 78

12. $2\frac{3}{4} - 1\frac{1}{2}$ $1\frac{1}{4}$

13. ${}^-17.1 + {}^-6.8$ ${}^-23.9$

14. ${}^-73.7 + {}^-24.1$ ${}^-97.8$

15. ${}^-15.3 - {}^-8.4$ ${}^-6.9$

16. $45.1 + {}^-3.5$ 41.6

17. ${}^-10 - 14.6$ ${}^-24.6$

18. ${}^-1\frac{1}{2} - {}^-2\frac{3}{10}$ $\frac{8}{10}$ or $\frac{4}{5}$

Mixed Applications

19. Melody bought 10 shares of ABX stock at $34\frac{1}{2}$ per share. One week later it fell to $31\frac{3}{8}$. After two weeks it was at $37\frac{1}{2}$. How much would she have lost per share if she had sold it after one week? How much would she have made per share if she had sold it after two weeks?

$3.125; $3.00

20. Benjamin purchased 100 shares of Hanover stock at $56\frac{1}{8}$ per share. It later went up to $60\frac{1}{2}$ and then down to $52\frac{1}{4}$. How much money would he have made per share if he had sold it at the higher price? How much money would he have lost if he had sold it at the lower price?

$4.375; $387.50

21. The security expense of a town budget is 25% of the entire budget. What fraction expresses the part of the town budget that is not security expense?

$\frac{3}{4}$

22. The perimeter of a rectangular field is 60 m. What are three possible measurements of the field? Use dimensions to the nearest meter. **Answers will vary.**

Possible answers: 15 m × 15 m, 20 m × 10 m, 12 m × 18 m

Name ______________________________

LESSON 5.5

Multiplying and Dividing Rational Numbers

Show the key sequence you would use to solve the problem with a calculator. **Possible answers are given.**

1. $^{-}7.6 \times 4$

7.6 [+/−] [×] 4 [=] $^{-}30.4$

2. $^{-}18.6 \div 3$

18.6 [+/−] [÷] 3 [=] $^{-}6.2$

3. $1\frac{1}{2} \times {}^{-}2\frac{1}{4}$

1.5 [×] 2.25 [+/−] [=] $^{-}3.375$

4. $^{-}3.5 \div {}^{-}7$

3.5 [+/−] [÷] 7 [+/−] [=] 0.5

Find the product or quotient.

5. $^{-}119 \div 7$ — $^{-}17$

6. $19 \times {}^{-}9$ — $^{-}171$

7. $^{-}182 \div {}^{-}14$ — 13

8. 23×15.4 — 354.2

9. $^{-}76.5 \div 17$ — $^{-}4.5$

10. $^{-}172.5 \div {}^{-}2.3$ — 75

11. $^{-}1\frac{1}{8} \div 6\frac{3}{4}$ — $\frac{-1}{6}$

12. $^{-}8\frac{1}{3} \times 1\frac{1}{5}$ — $^{-}10$

Find the quotient, and tell what two integers the quotient is between.

13. $1{,}450 \div 95$ — 15.26; between 15 and 16

14. $31{,}593 \div {}^{-}124$ — $^{-}254.78$; between $^{-}254$ and $^{-}255$

15. $^{-}45{,}671 \div 340$ — $^{-}134.33$; between $^{-}134$ and $^{-}135$

16. $^{-}187{,}302 \div 3{,}700$ — $^{-}50.62$; between $^{-}50$ and $^{-}51$

17. $^{-}203{,}426 \div {}^{-}1{,}230$ — 165.39; between 165 and 166

Mixed Applications

18. Diana saves $5.25 each week. How much will she save in a year? There are 52 weeks in a year.

$273

19. Jocelin has read 20% of the book she selected for a report. What fraction of the book has she read?

$\frac{1}{5}$ of the book

20. In 6 months, Cooper will need $751.20 for an airline ticket. How much should he save each month so that he will have enough money to buy it?

$125.20

21. Everett left home at 9:00 A.M. He walked for 45 min, rested for 35 min, then walked for 20 min, and arrived at Marshall's house. What time did he arrive at Marshall's house?

10:40 A.M.

Use with text pages 110–111.

Name ______________________

Numerical and Algebraic Expressions

Vocabulary

Write the correct letter from Column 2.

1. numerical expression ___b___
2. algebraic expression ___c___
3. variable ___a___

a. a letter or symbol that stands for one or more numbers
b. a phrase that has only numbers and operation symbols
c. a phrase that includes variables, numbers, and operation symbols

---

Match the correct algebraic expression to the words.

$\frac{3}{n}$
$\frac{n}{3}$
$n + 3$

4. 3 more than a number, n ___$n + 3$___
5. 3 divided by a number, n ___$\frac{3}{n}$___

Write a numerical expression for each word expression.

6. 3 less than the product of 4 and 6

   ___$4 \times 6 - 3$___

7. 2 more than the number 5 squared

   ___$5^2 + 2$___

Write an algebraic expression for each word expression. **Expressions may vary.**

8. twice the length, l, times the width, w

   ___$2lw$___

9. 7 times the sum of 5 and a number, k

   ___$7(5 + k)$___

Write a word expression for each algebraic expression. **Expressions may vary.**

10. $2(m + 4)$

    ___twice the sum of m and 4___

11. $a - 10$

    ___10 less than a___

Mixed Applications

12. The seventh-grade class raised $245 to spend on books. They purchase n books, each costing the same amount. Write an expression that represents the cost of each book.

    ___$245 \div n$___

13. The temperature of a chemical reaction is $^-55.6$°C. After 50 minutes, it changes by $^-7.4$°C. What is the new temperature?

    ___$^-63$°C___

Use with text pages 120–121.

Name ______________________

LESSON 6.2

Evaluating Expressions

Vocabulary

Complete.

1. When you **evaluate** a numerical expression, you put it in its simplest numerical form, as a single number.

Tell what operation you would perform first.

2. $5 \times 4 - 8$ **multiplication**

3. $(7 - 2) \div 5$ **subtraction**

4. $4 \div 2 \times 12$ **division**

Evaluate each expression.

5. $14 \div 2 + 3 \times 5$ **22**

6. $5 \times 6 - 3^3$ **3**

7. $45 \div (^-9) - 7\frac{1}{2}$ **$^-12\frac{1}{2}$**

Evaluate the expression $2.5n + 4.5$ for each value of n.

8. $n = 2$ **9.5**

9. $n = {^-1}$ **2**

10. $n = 6$ **19.5**

Evaluate the expression for the given values of the variables.

11. $3a - 5b$ for $a = 3$ and $b = {^-4}$ **29**

12. $3.5s + (s - t)$ for $s = 3$ and $t = 0.5$ **13**

13. $\frac{m}{2} - 2n$ for $m = 3$ and $n = {^-1.5}$ **4.5**

Mixed Applications

14. Cara bought g containers of juice that cost \$0.55 each. Write an expression that represents the total cost of the juice. Then find the total cost if $g = 24$.

    $0.55g$; \$13.20

15. At Grace's Grill, the price of yogurt is $\frac{1}{5}$ the price, s, of a chef salad. Write an expression that represents the cost of yogurt. Then find the cost of yogurt if $s = \$3.50$.

    $\frac{1}{5}s$; \$0.70

16. The average of 3 numbers is 6.5. Two of the numbers are 7.1 and 5.4. What is the third number?

    7

17. Kevin gives one half of his baseball cards to Fernando. Then he gives one half of what is left to Jamie. Then he gives one third of what is left to Pat. If he has 8 cards left, how many did he start with?

    48 cards

Use with text pages 122–125.

Name ______________________

LESSON 6.3

Combining Like Terms

Vocabulary

Complete.

1. When you combine like terms in an expression, you are putting the expression in **simplest form**.

Identify the like terms in each list of terms.

2. $3g, 8h, 7, 4g$ — **$3g$ and $4g$**

3. $2y, 18, 3y^2$ — **no like terms**

4. $4, 2a, 3b, 5a, 3a^2$ — **$2a$ and $5a$**

Combine like terms.

5. $8y + 3 + 2y$ — **$10y + 3$**

6. $5q - 4 - 6q$ — **$^-q - 4$**

7. $2x + 2 + 3x - x$ — **$4x + 2$**

Combine like terms. Then evaluate each expression for $x = 8$.

8. $7x - 2.5x + 4$ — **$4.5x + 4$; 40**

9. $4.5x + 17 + 2x$ — **$6.5x + 17$; 69**

10. $2(x + 3) - 5x$ — **$^-3x + 6$; $^-18$**

Mixed Applications

11. Mitchell works for m dollars per hour. He worked 5 hr Friday and 6 hr Saturday. Write an expression for the total amount Mitchell earned. Then evaluate the expression for $m = \$5.50$.

$5m + 6m$; $11m$; \$60.50

12. Yolanda saves \$3.25 per week for w weeks. Then she saves \$2.50 per week for w more weeks. Write an expression for the amount she saves. Evaluate the expression for $w = 25$.

$3.25w + 2.50w$; $5.75w$; \$143.75

13. A sector of a circle graph represents 40% of XYZ Company's revenues. What is the measure of the central angle of this sector?

144°

14. Kendra receives a grade of 86% on a history test. There are 50 questions on the test. How many questions did Kendra answer correctly?

43 questions

Name ______________________

LESSON 6.4

Sequences and Expressions

Write the correct letter from Column 2.

1. arithmetic sequence ___c___
2. term ___a___
3. common difference ___b___

a. an element, or a number, in a sequence
b. the difference between any two successive terms
c. a sequence in which the difference between any term and the one after it is always the same

Tell whether the sequence is an arithmetic sequence. If it is, find the common difference.

4. 0, 5, 10, 15, 20, . . . ___yes___
5. 1, 1, 2, 2, 3, 3, . . . ___no___
6. 1, 2, 5, 6, 9, 10, . . . ___no___

Find the common difference for each sequence.

7. 3, 9, 15, 21, 27, . . . ___6___
8. 94, 89, 84, 79, . . . ___$^{-}5$___
9. 17, 20, 23, 26, . . . ___3___

Write the next three terms in each sequence.

10. 14, 28, 42, 56, . . . ___70, 84, 98___
11. 6, 3, 0, $^{-}3$, . . . ___$^{-}6$, $^{-}9$, $^{-}12$___
12. 4, 9, 14, 19, . . . ___24, 29, 34___

Write an expression to describe each sequence. Then use the expression to find the tenth term.

13. 15, 20, 25, 30, . . . ___$15 + 5(n - 1)$; 60___
14. 10, 9, 8, 7, . . . ___$10 + {}^{-}1(n - 1)$; 1___
15. 1, 3, 5, 7, . . . ___$1 + 2(n - 1)$; 19___

Mixed Applications

16. Mrs. Shin has $1,200 in the bank. She earns $1 each week in interest. Write an expression to describe the pattern. Then tell how much money she has after 52 weeks.

___$1{,}200 + (1 \times n)$; $1,252___

17. Bert the lion weighed 1,600 lb. Then he got sick and started to lose 10 lb per day. Write an expression to describe the pattern and tell what day Bert weighed 1,490 lb.

___$1{,}600 - 10d$; 11th day___

18. The temperature at Glens Falls is 12°F at 6 A.M. At noon the temperature is $^{-}3$°F. What is the average hourly change?

___$^{-}2.5$°F___

19. There are n questions on a language test. Keith answered 18 correctly. Write an expression that represents his grade. Then find his grade as a percent for $n = 24$.

___$\frac{18}{n}$; 75%___

Name ______________________

Connecting Equations and Words

Vocabulary

1. A(n) equation is a sentence that shows two expressions are equivalent.

Choose the equation that represents the problem.

2. Ross has 3 containers of marbles. The same number of marbles is in each container. Ross has a total of 51 marbles. How many are in each container? b

   a. $3 + 51 = m$ b. $\frac{51}{3} = m$ c. $\frac{3}{51} = m$

3. Marsha buys a new pair of jeans. The jeans cost 3 times the amount of money she has saved. If the jeans cost $24, how much has Marsha saved? c

   a. $3 + 8 = s$ b. $24 - s = 3$ c. $3s = 24$

First, choose a variable, and tell what it represents. Then, write an equation for each word sentence. **Variables and equations may vary.**

4. Twelve less than the number of cans collected is 35.

   n = cans collected; $n - 12 = 35$

5. Five times the cost of a book is $19.75.

   b = cost of book; $5b = 19.75$

Mixed Applications

For Problems 6–8, write the equation you would use to solve the problem. Do not solve. **Variables and equations may vary.**

6. Jill and her four friends bought a present for their coach. The total cost of the gift was divided equally among them. Each paid $3.35. How much did the gift cost?

   $x \div 5 = 3.35$

7. Bill puts $12.50 in his bank account each month. He now has $217 in the account. For how many months has Bill been saving?

   $12.50x = 217$

8. Max spent $5.98 less than Rod on school supplies. Max spent $19.88. How much did Rod spend?

   $x - 5.98 = 19.88$

9. The distance to the sun is approximately 93,000,000 mi. Express the number of miles in scientific notation.

   9.3×10^7 mi

Name ______________________

LESSON 7.2

Solving Addition and Subtraction Equations

Tell whether the given value is the solution of the equation. Write *yes* or *no*.

1. $b - 17 = 36, b = 13$ — no
2. $r + 26 = 54, r = 29$ — no
3. $m + 42 = 71, m = 29$ — yes
4. $s - 66 = 15, s = 81$ — yes
5. $48 = n + 29, n = 19$ — yes
6. $117 = c - 54, c = 63$ — no

Tell whether you would add or subtract to solve.

7. $m - 3.4 = 1.8$ — add
8. $14 + b = 23.3$ — subtract
9. $x + 3\frac{1}{4} = 5\frac{3}{8}$ — subtract

Solve and check.

10. $2.86 + t = 6$ — $t = 3.14$
11. $b + 15 = 59$ — $b = 44$
12. $c - 34 = {}^-25$ — $c = 9$
13. $16\frac{1}{2} = r + 5$ — $11\frac{1}{2} = r$
14. $4\frac{2}{5} = n + 2\frac{3}{10}$ — $n = 2\frac{1}{10}$
15. $z - 1\frac{3}{4} = 7\frac{1}{2}$ — $z = 9\frac{1}{4}$
16. $y + \frac{3}{4} = 1\frac{1}{2}$ — $y = \frac{3}{4}$
17. $0.4 + q = 1.6$ — $q = 1.2$
18. $4 + k = {}^-10$ — $k = {}^-14$

Mixed Applications

For Problems 19–20, choose a variable and write an equation. Then find the value of the variable. **Variables and equations will vary.**

19. Twelve points less than the number of points scored is 53 points.

    p = points; $p - 12 = 53$; $p = 65$

20. The number of inches, when increased by 12.5 inches, becomes 40 inches.

    x = inches; $x + 12.5 = 40$;
    $x = 27.5$

21. Ruth paid $32.75 for jeans and a shirt. The shirt cost $12.95. How much did the jeans cost?

    $19.80

22. Max purchased 54 yd^2 of flooring. The total cost of the flooring was $429.30. What was the cost per square yard?

    $7.95

Use with text pages 141–143.

Name ____________________

LESSON 7.3

Multiplication and Division Equations

Tell whether the given value is the solution of the equation. Write *yes* or *no*. If the value is not the solution, solve the equation.

1. $5x = {}^{-}5, x = 1$
   no; $x = {}^{-}1$
2. $\frac{m}{1.4} = 5, m = 7$
   yes
3. $0.3r = 0.12, r = 4$
   no; $r = 0.4$
4. $15t = 60, t = 4$
   yes
5. $\frac{m}{8} = 3.2, m = 24$
   no; $m = 25.6$
6. $\frac{s}{0.5} = 0.5, s = 0.25$
   yes

Solve and check.

7. $\frac{m}{5} = 17$
   $m = 85$
8. $9b = {}^{-}108$
   $b = {}^{-}12$
9. $5.4 = \frac{b}{4}$
   $b = 21.6$
10. $\frac{y}{2.3} = 18.5$
    $y = 42.55$
11. $8h = 36.8$
    $h = 4.6$
12. $0.4b = 24$
    $b = 60$
13. $28z = 56$
    $z = 2$
14. $\frac{q}{{}^{-}2} = 8$
    $q = {}^{-}16$
15. $8r = {}^{-}64$
    $r = {}^{-}8$

For each word sentence, choose a variable and write an equation. Then find the value of the variable. **Variables will vary.**

16. The cost of a stereo system is divided into 12 equal payments of $47.50.
    $\frac{s}{12} = 47.50; s = 570$
17. A group of people renting a hall pay $18 each, for a total of $450.
    $18p = 450; p = 25$

Mixed Applications

The formula $d = rt$ relates distance (d), rate (r), and time (t). Use this formula to solve Problems 18 and 19.

18. A jet flew at an average rate of 450 mi per hr. If the flight lasted $3\frac{1}{2}$ hr, how far did the plane travel?
    1,575 mi
19. Jerome bicycled 91 mi. If he rode for $6\frac{1}{2}$ hr, what was his average rate in miles per hour?
    14 miles per hour
20. Mike's bowling score is 17 points less than Lisa's. If Mike's score is 121, what is Lisa's score?
    138 points
21. Room and board is 45% of Tino's budget. What is the central angle measure of "Room and Board" on a circle graph of Tino's budget?
    162°

Name ______________________

LESSON 7.4

Problem-Solving Strategy

Working Backward to Solve Problems

Work backward to solve each problem.

1. Ron finished work at 3:45 P.M. He had been working for $7\frac{1}{2}$ hr. When did Ron start work?

   8:15 A.M.

2. Lin spent $5.18 more than Tim at the mall. Lin spent $26.98. How much did Tim spend?

   $21.80

3. Mrs. Rose has a teacup collection. She has a total of 17 cups in her collection. Each year for the past 3 years, Mrs. Rose has added 4 cups to the collection. How many cups did she have originally?

   5 cups

4. Michele receives a monthly allowance. This month she put half of her allowance in the bank. Then she bought school supplies for $9.27. She has $5.73 left. What is Michele's monthly allowance?

   $30.00

Mixed Applications

Solve.

CHOOSE A STRATEGY

- Write an Equation
- Use a Formula
- Guess and Check
- Make a List
- Work Backward

Choices of strategies will vary.

5. A cyclist traveled at an average speed of 16 mph. If the cyclist rode for $5\frac{1}{4}$ hr, how far did she go?

   84 mi

6. At a carnival, Zoe spent $12.38 less than Zack. Zoe spent $18.52. How much did Zack spend?

   $30.90

7. For her parade costume, Beth has a choice of 4 skirts—red, green, blue, and white—and 3 shirts—pink, yellow, and orange. Make a list to help you find the number of possible skirt-and-shirt combinations.

   RP, RY, RO, GP, GY, GO, BP, BY,
   BO, WP, WY, WO; 12 combinations

8. Jack put $\frac{1}{3}$ of his paycheck in the bank. Then he spent $29 on tapes and CDs. He has $35 left. What was the total of Jack's paycheck?

   $96

9. A theater group has 226 members. There are 42 more women than men in the group. How many men are in the group?

   92 men

10. Max has as many nickels as quarters. The total value of the coins is $2.40. How many quarters does Max have?

    8 quarters

Use with text pages 146–147.

Name ______

LESSON 7.5

Proportions

Vocabulary

1. A(n) **proportion** is a special kind of equation that states that two ratios are equivalent.

Use a model to solve each proportion. Draw a diagram of your model and solution. **Check students' models.**

2. $\frac{1}{3} = \frac{n}{36}$ **$n = 12$**

3. $\frac{3}{8} = \frac{n}{32}$ **$n = 12$**

Solve each proportion.

4. $3:15 = b:5$ **$b = 1$**

5. $2:p = 12:36$ **$p = 6$**

6. $7:11 = s:44$ **$s = 28$**

7. $\frac{a}{27} = \frac{18}{81}$ **$a = 6$**

8. $\frac{54}{42} = \frac{t}{14}$ **$t = 18$**

9. $\frac{2.5}{7} = \frac{x}{3.5}$ **$x = 1.25$**

10. $\frac{m}{19} = \frac{11}{95}$ **$m = 2.2$**

11. $\frac{3.2}{4.6} = \frac{r}{4.6}$ **$r = 3.2$**

12. $\frac{j}{8} = \frac{1.4}{4}$ **$j = 2.8$**

Solve each proportion for d. Let $a = 3$, $b = 4$, and $c = 12$.

13. $\frac{a}{b} = \frac{c}{d}$ **$d = 16$**

14. $\frac{3a}{2b} = \frac{d}{c}$ **$d = 13.5$**

15. $\frac{ab}{a+c} = \frac{ac}{d}$ **$d = 45$**

Mixed Applications

16. During a special sale, Vic's Music World sold 92 CDs in 0.75 hr. At this rate, how many CDs would be sold in 3 hr?

368 CDs

17. Rehearsal for the school play ended at 4:35 P.M., after $3\frac{1}{4}$ hours. When did the rehearsal begin?

1:20 P.M.

18. A member of the school track team runs the 300-m dash in 51 sec. At the same rate, how long would it take her to run 400 m?

68 sec

19. On payday Millie put $\frac{1}{2}$ of her paycheck in a savings account. She then spent $47.65 on groceries and $12.00 on gasoline. She has $40.35 left. What was the amount of Millie's paycheck?

$200

Name ________________________________

LESSON 8.1

Problem-Solving Strategy

Write an Equation to Solve Two-Step Problems

Write an equation and solve. Variables may vary.

1. The number of base hits in softball that Ryan has is 5 fewer than twice the number of hits that Erica has. If Ryan has 17 base hits, how many hits does Erica have?

   $2n - 5 = 17$; $n = 11$;

   11 base hits

2. This week Rhea earned $5.25 more than double the money she earned last week. If Rhea earned $36.75 this week, how much did she earn last week?

   $2n + 5.25 = 36.75$;

   $n = 15.75$; $15.75

3. The length of a school lunchroom is 15 ft longer than 2 times its width. If the length is 135 ft, what is the width?

   $2x + 15 = 135$;

   $x = 60$; 60 ft

4. This month Julie delivered 18 fewer than 3 times as many newspapers as she did last month. She delivered 168 newspapers this month. How many did she deliver last month?

   $3n - 18 = 168$; $n = 62$;

   62 newspapers

Mixed Applications

Solve.

CHOOSE A STRATEGY

- Write an Equation
- Guess and Check
- Work Backward
- Draw a Diagram
- Use a Formula

Choices of strategies will vary.

5. Five members of a baseball team line up for batting practice. The pitcher stands in front of the catcher. The right fielder stands behind the left fielder. The shortstop stands between the pitcher and the right fielder. In what order will they bat?

   left fielder, right fielder,

   shortstop, pitcher, catcher

6. There are 116 students in the seventh grade. The number of students who walk to school is 3 times as many as the number of students who ride a bus to school. How many seventh-grade students ride a bus to school?

   29 students

7. The Garden Club is planting rosebushes around Town Square. Town Square is 70 ft long and 55 ft wide. If the bushes are planted 5 ft apart, how many bushes does the club need?

   50 bushes

8. You can use the formula $F = \frac{9}{5}C + 32$ to convert a temperature expressed in degrees Celsius (C) to a temperature expressed in degrees Fahrenheit (F). Convert 21°C to degrees Fahrenheit.

   69.8° F

Use with text pages 158–159.

Name ______________________

LESSON 8.2

Simplifying and Solving

Tell whether the given value is a solution to the equation. Write *yes* or *no.*

1. $4s - 10 = 22; s = 8$ ___yes___

2. $9t + 5 = 68; t = 7$ ___yes___

3. $20x - 3x = 34; x = 3$ ___no___

4. $7a + 2(13 - a) = 196; a = 34$ ___yes___

5. $7(8 - n) + 8n = 57; n = 2$ ___no___

6. $21m - 13m + 15 = 31; m = 2$ ___yes___

Solve and check.

7. $5b - 7 + 2b = 56$ ___$b = 9$___

8. $17r + 6 - 8r = 105$ ___$r = 11$___

9. $2(b - 8) + 45 = 81$ ___$b = 26$___

10. $3^2 + 17.2g - 3.6g = 63.4$ ___$g = 4$___

11. $5m + 2^3 + 7 = 115$ ___$m = 20$___

12. $96 = 6(t - 5) + 8t$ ___$t = 9$___

Mixed Applications

Write an equation and solve.

13. The interest on Jon's savings account this year is $12.73 less than the amount of interest he received last year. The total interest for the two years is $64.31. How much interest did he earn this year?

$i + (i - 12.73) = 64.31$;

$25.79 this year

14. A 52-in. piece of wire is cut into three pieces. One piece is twice as long as each of the other two. What is the length of the longest piece of wire?

$2w + w + w = 52$;

longest piece is 26 in.

15. This week Josh earned $6.32 less than twice the amount he earned last week. If he earned $59.58 this week, how much did Josh earn last week?

$2w - 6.32 = 59.58$; $32.95

16. The length of a vegetable garden is 6.5 ft longer than twice its width. If the length is 49.5 ft, what is the width?

$2w + 6.5 = 49.5$; 21.5 ft

Name ______________________

LESSON 8.3

Comparing Equations and Inequalities

Vocabulary

Complete.

1. A(n) ___inequality___ is a number sentence that shows the order relationship between two quantities that are not equal.

Replace the variable with each given value. Tell whether the algebraic sentence is an *equation* or an *inequality.*

$m + 7 \bigcirc 12$

2. $m = 0$ ___inequality___
3. $m = 9$ ___inequality___
4. $m = 5$ ___equation___

$6b - 5 \bigcirc 19$

5. $b = 4$ ___equation___
6. $b = 2$ ___inequality___
7. $b = 7$ ___inequality___

Tell whether x is a solution to the inequality. Write *yes* or *no.*

8. $3x - 5 \geq 3; x = 2$ ___no___
9. $x - 3.2 + 2^3 < 5.1; x = .2$ ___yes___

Write an inequality or an equation for each graph. **Possible answers are given.**

10.

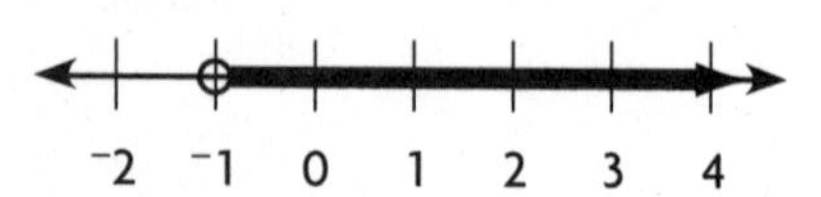

___$x > {}^-1$___

11.

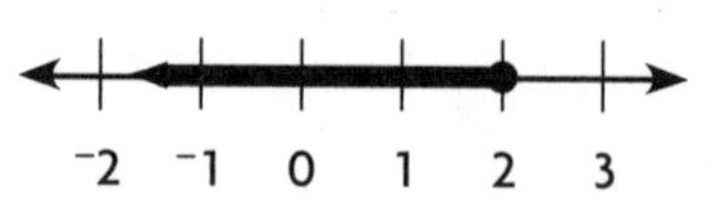

___$x \leq 2$___

Mixed Applications

12. Pedro earned $25 more this week on his paper route than he did last week. This week he earned more than $60. Write an inequality for these word sentences. Let *e* represent the amount he earned last week.

___$e + 25 > 60$___

13. Rick's card collection contains 11 less than 3 times the number of cards that Chet has. If Rick has 70 cards, how many does Chet have?

___27 cards___

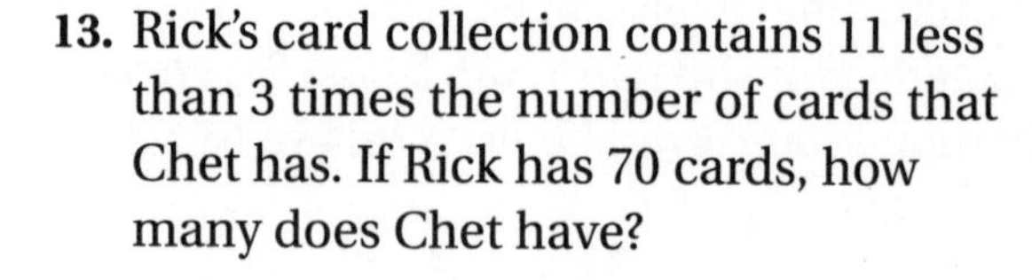

14. For Problem 12, write an inequality to describe the numbers that are not part of the solution.

___Possible answer: $x > 2$___

15. Gus bought 6 notebooks, a pen for $1.19, and lined paper for $2.09. He spent a total of $8.02. How much did 1 notebook cost?

___$0.79___

Use with text pages 163–165.

Name ______________________________

Solving Inequalities

Tell what operation and number you would use to solve each inequality.

1. $p - 4.6 \leq 2$ **add; 4.6**
2. $0.4s > 0.004$ **divide; 0.4**
3. $a - 3\frac{1}{8} \neq 2$ **add; $3\frac{1}{8}$**
4. $a + 4 \geq 12$ **subtract; 4**
5. $\frac{x}{50} \neq 1.3$ **multiply; 50**
6. $3^4 < \frac{m}{5}$ **multiply; 5**

Solve. Write the whole numbers that make the inequality true.

7. $y - 5 \geq 2$ **$y \geq 7$; 7, 8, 9, . . .**
8. $3m < 9$ **$m < 3$; 0, 1, 2**
9. $2n + 3 > 17$ **$n > 7$; 8, 9, 10, . . .**

Solve the equation or inequality.

10. $^-2x = 16$ **$x = {^-8}$**
11. $2y + 3 \neq 15$ **$y \neq 6$**
12. $\frac{s}{3} \leq {^-4}$ **$s \leq {^-12}$**
13. $q + 5 \geq 7$ **$q \geq 2$**
14. $8 + z < 15$ **$z < 7$**
15. $m - 14 = 2$ **$m = 16$**

Solve the equation or inequality. Graph the solution. **Check students' graphs.**

16. $4m > 20$ **$m > 5$**
17. $6b = 15$ **$b = 2\frac{1}{2}$**
18. $x + 3 \leq 2$ **$x \leq {^-1}$**
19. $5b - 4 \leq 31$ **$b \leq 7$**
20. $0.3z \geq 1.8$ **$z \geq 6$**
21. $\frac{r}{2} = 3\frac{1}{2}$ **$r = 7$**

Mixed Applications

22. Huntsville Soccer Club bought new goals and flags. The goals cost 5 times as much as the flags. The goals cost less than $250. Is it possible that the flags cost $45?

    $f < 50$; yes

23. A total of 24 books are stacked into 3 piles. One pile contains twice as many books as are in each of the other two. How many books are in each pile?

    12 books; 6 books; 6 books

24. Mia has bowled 4 out of 5 games. Her scores are 102, 88, 114, and 96. Mia wants her average score for 5 games to be greater than 100. What is the lowest score Mia can get in her fifth game?

    $x > 100$; 101

25. Ron wants to buy a camera that costs $189. He earns $55 every two weeks for delivering newspapers. Ron decides to save all his earnings for the camera. How long will it take him to save enough?

    4 pay periods, or 8 weeks

Use with text pages 166–167.

Name ______________________

LESSON 9.1

Graphing Ordered Pairs

Vocabulary

Write the correct letter from column 2.

1. first number of an ordered pair __d__
2. coordinate plane is divided into these __c__
3. where x-axis and y-axis intersect __a__
4. second number of an ordered pair __e__
5. ordered pairs name the positions of these __b__

a. origin
b. points
c. quadrants
d. *x*-coordinate
e. *y*-coordinate

Write the ordered pair for each point.

6. *A* __(⁻2,2)__ 7. *B* __(2,6)__ 8. *C* __(4,⁻2)__

9. *D* __(⁻1,⁻1)__ 10. *E* __(⁻4,⁻5)__ 11. *F* __(2,2)__

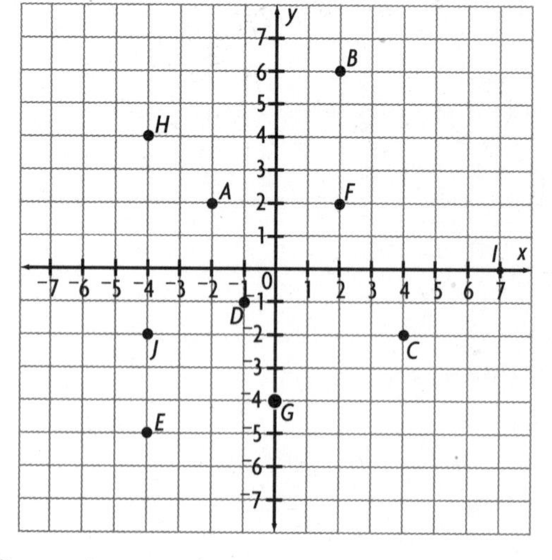

Write a letter for each ordered pair.

12. (7,0) __*I*__ 13. (⁻4,4) __*H*__ 14. (⁻4,⁻2) __*J*__

15. Are (⁻4,4) and (⁻4,⁻2) located in the same quadrant? Explain.

No; (⁻4,4): Quadrant II; (⁻4,⁻2): Quadrant III

16. In which quadrant is a point located where the *x*-coordinate is negative and the *y*-coordinate is positive? where the *x*-coordinate is positive and the *y*-coordinate is negative?

Quadrant II; Quadrant IV

Mixed Applications

17. Ryan and Beth start hiking from the same point. Ryan hikes 5 mi south and then 4 mi west. Beth hikes 2 mi east and then 5 mi south. How far apart are they at the end of their hikes?

6 mi

18. Brian bought some CDs and tapes. The CDs cost exactly three times as much as the tapes. The CDs cost $27. Is it possible that the tapes cost $12? Explain.

no; $3t = 27$; $t = 9$; $9 \neq 12$

Use with text pages 172–173.

Name ______________________

LESSON 9.2

Relations

Vocabulary

Complete.

1. The numbers, words, or objects in a set are **elements** of the set.

2. You form a(n) **relation** when you match the elements of one set to the elements of another set.

3. The first set of elements in a relation is the **domain**.

4. The second set of elements in a relation is the **range**.

Write the ordered pairs for the relation.

5.

Ted's age, t	1	2	3	4	5
Don's age, d	15	16	17	18	19

(1,15), (2,16), (3,17), (4,18), (5,19)

Write an equation for the relation.

6.

Sale price, r	\$4	\$6	\$8	\$10
Original price, p	\$8	\$12	\$16	\$20

$p = 2r$

Use the relation {(⁻3,0), (⁻2,1), (⁻1,2), (0,3), (1,4), (2,5)} for Exercises 7–8.

7. What is the domain of the relation? What is the range of the relation?

{⁻3, ⁻2, ⁻1, 0, 1, 2}; {0, 1, 2, 3, 4, 5}

8. Write an equation that represents the relation.

$y = x + 3$

9. Complete the table for the relation, $y = x + {}^{-}2$. Then graph the relation.

x	⁻3	0	1	3
y	⁻5	⁻2	⁻1	1

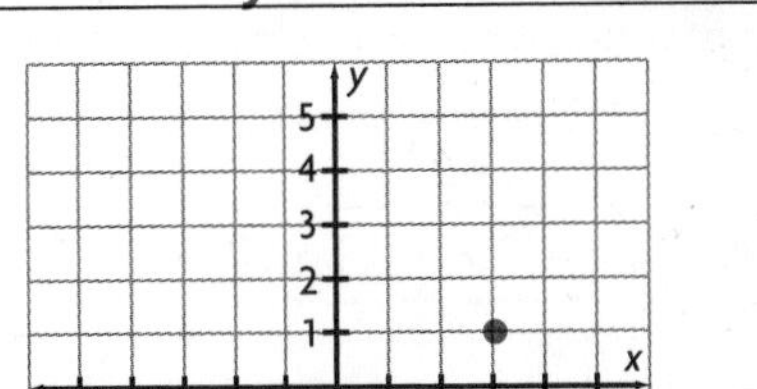

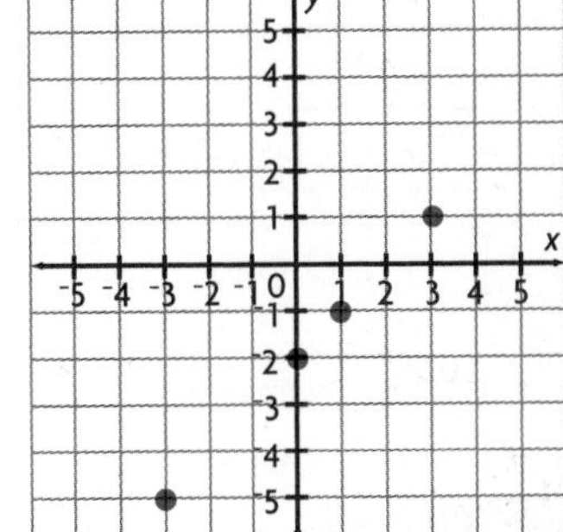

Mixed Applications

10. Tom is driving his motorcycle at an average speed of 52 miles per hour. Write an equation for the relation between the number of hours, x, and the distance in miles, y, that he travels.

$y = 52x$

11. Rhonda and Zack start walking from the same point. Rhonda walks 8 miles east, then 2 miles north. Zack walks 2 miles west, then 2 miles north. How far apart will they be when they stop?

10 mi

Name ______________________________

Functions

Vocabulary

Complete.

1. A(n) ___function___ is a relation in which each element of the domain corresponds to one and only one member of the range.

2. You use the ___vertical line test___ to see if a vertical line crosses two or more points on a graph.

Is the relation a function? Write *yes* or *no.* If you write *no,* explain.

3. {(1,3), (2,3), (3,7), (4,9)}

___yes___

4. {(⁻1,⁻2), (1,3), (⁻1,3), (3,1)}

___No; ⁻1 in the domain is repeated.___

5. {(0,5), (⁻1,6), (6,⁻1), (5,0)}

___yes___

Use the vertical line test to determine if the relation is a function.

6. 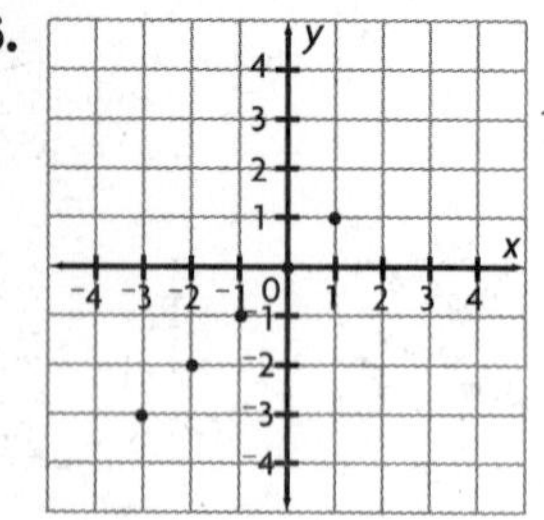

___yes___

7. 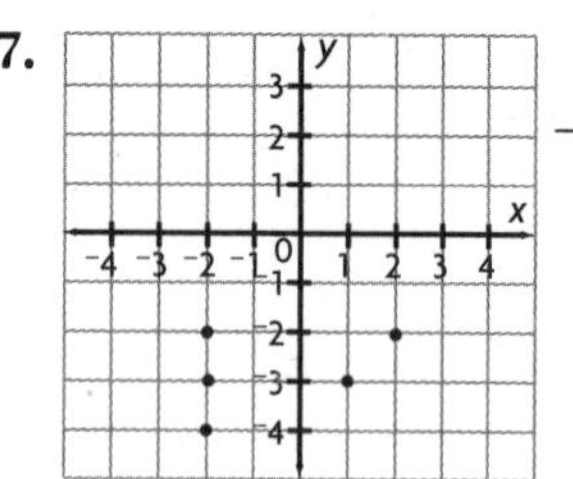

___no___

8. 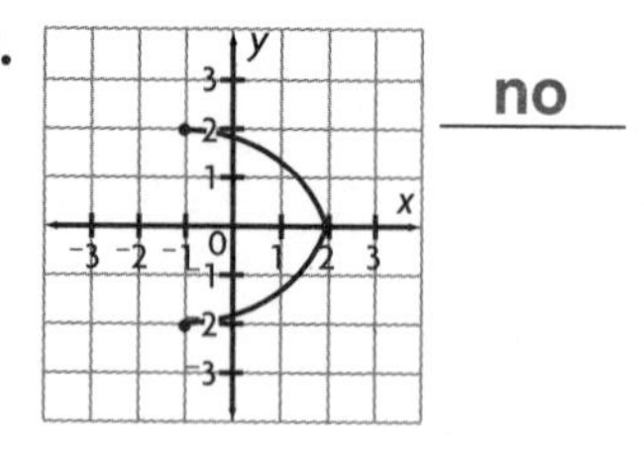

___no___

Write the ordered pairs for the relation.
Then graph the set of ordered pairs.

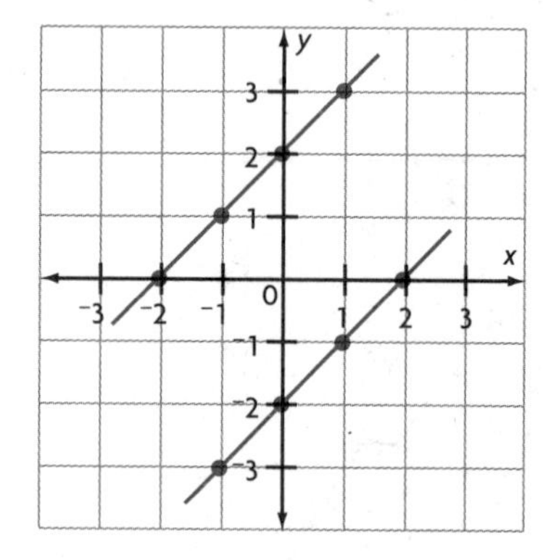

9.

x	⁻1	0	1	2
y	⁻3	⁻2	⁻1	0

___(⁻1,⁻3), (0,⁻2), (1,⁻1), (2,0)___

10.

x	⁻2	⁻1	0	1
y	0	1	2	3

___(⁻2,0) (⁻1,1) (0,2) (1,3)___

Mixed Applications

11. Write an equation for the sentence *Five less than x equals y.* Choose three values of *x*, and find the matching values of *y.* Does the equation represent a function?

___$y = x - 5$; (0,⁻5), (1,⁻4), (5,0); yes___

12. Anita bought a new bicycle. She rode 5^2 miles the first month and 5^3 miles the second month. How much farther did she ride the second month?

___100 mi___

Use with text pages 177–179.

Name ______________________________

Linear Equations

Vocabulary

Complete.

1. When you graph ordered pairs and the graph is a straight line, you have graphed a(n) __linear equation__.

For each equation, replace x with 2. Write the solution as an ordered pair.

2. $y = x + 2$

__(2,4)__

3. $y = ^{-}3x$

__(2,⁻6)__

4. $y = \frac{1}{2}x$

__(2,1)__

5. $^{-}2x = y$

__(2,⁻4)__

Make a table of values for each equation. Write the ordered pairs.
Check students' tables. Answers may vary; possible answers are given.

6. $y = 3x - 2$

__(1,1), (2,4), (⁻1,⁻5)__

7. $y = ^{-}2x + 3$

__(0,3), (1,1), (2,⁻1)__

Write the equation for each table of values.

8.

x	y
⁻1	⁻2
1	0
2	1

__$y = x - 1$__

9.

x	y
⁻1	2
0	3
1	4

__$y = x + 3$__

10.

x	y
⁻1	1
0	0
1	⁻1

__$y = ^{-}x$__

Mixed Applications

11. Jill earns $4 per hour for baby sitting.

a. Make a table of values to show her hours and wages for 1 to 4 hours.

Hours, x	1	2	3	4
Wages, y	$4	$8	$12	$16

b. Write a linear equation for the function.

__$y = 4x$__

c. Graph the equation.

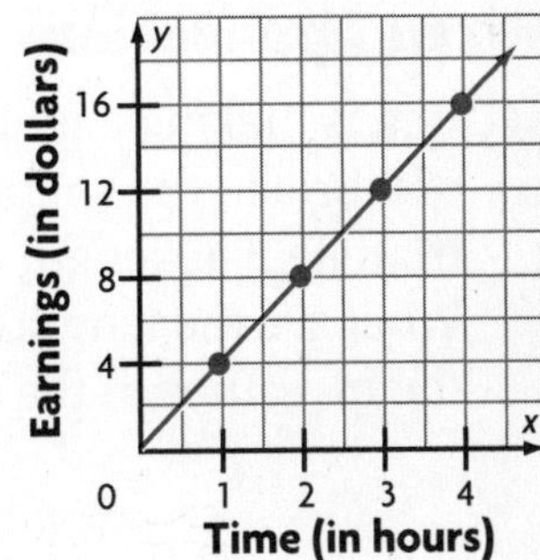

Name ______________________

Congruent Line Segments and Angles

Vocabulary

Complete.

1. Opposite angles formed when two lines intersect are **vertical angles**.

$\angle ABC$ is congruent to $\angle DBF$. Find the measure of each angle.

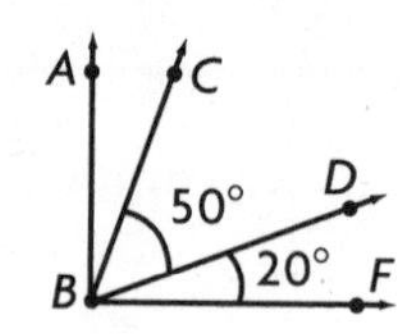

2. $\angle ABC$ **20°** 3. $\angle CBF$ **70°** 4. $\angle ABD$ **70°**

For Exercises 5 and 6, use the figure at the right.

R
W
S
T

5. Which are vertical angles? adjacent angles?

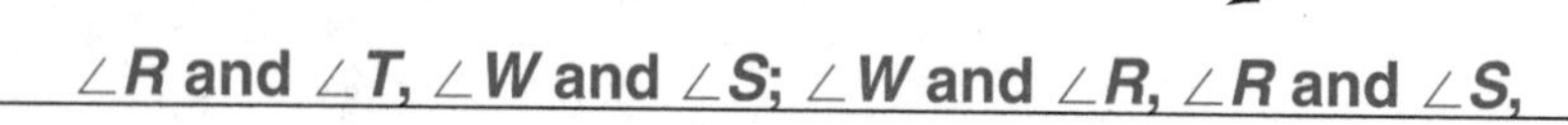

$\angle R$ and $\angle T$, $\angle W$ and $\angle S$; $\angle W$ and $\angle R$, $\angle R$ and $\angle S$,

$\angle S$ and $\angle T$, $\angle T$ and $\angle W$

6. Which angles are congruent?

$\angle R$ and $\angle T$, $\angle W$ and $\angle S$

Triangle *JKL* is an isosceles triangle. Find each measure.

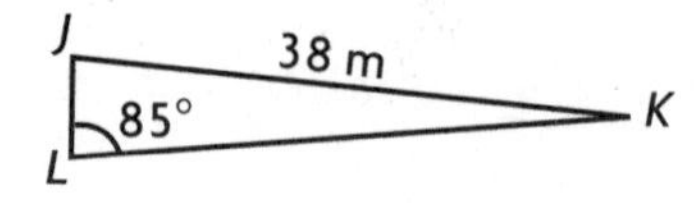

7. $\angle KJL$ **85°** 8. $\angle JKL$ **10°** 9. side KL **38m**

Write *always, sometimes,* or *never.*

10. An equilateral triangle has three congruent angles. **always**

11. Adjacent angles are not congruent. **sometimes**

12. One angle of an equilateral triangle measures 45°. **never**

Mixed Applications

13. Rita is making a quilt. She cuts a piece of fabric in the shape of an equilateral triangle. The length of one side is 3.5 in. What are the lengths of the other sides? the measures of the angles?

each side: 3.5 in.;

each angle: 60°

14. Of 750 students surveyed, 225 said they are members of a sports team. What percent of the students belong to a sports team?

30%

Use with text pages 193–196.

Name ______________________________

Symmetry

Vocabulary

Complete.

1. When a figure matches the original figure after a rotation of less than 360° about a central point, **rotational symmetry** exists.
2. A figure with rotational symmetry can be rotated about a central point, called the **point of rotation**.
3. When two halves of a figure can be made to match by folding on a line, **line symmetry** exists.
4. A figure with rotational symmetry will **coincide with, or match,** the original figure after being rotated less than 360°.

Write *line, rotational,* or *none* to describe the symmetry of each figure.

5. 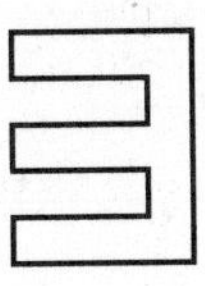

line

6. 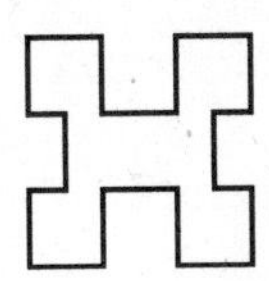

line, rotational

7. 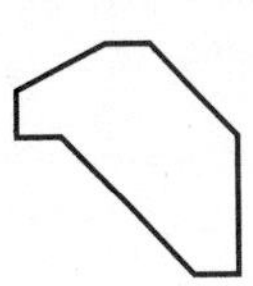

none

8. 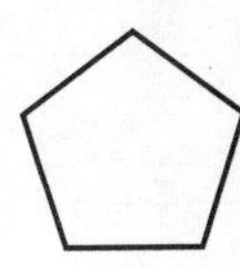

line, rotational

Identify the turn symmetry for each figure as a fraction of a turn and in degrees.

9. 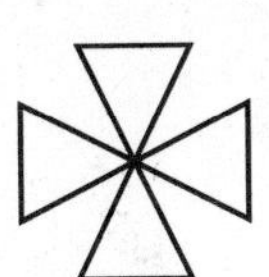

$\frac{1}{4}$ turn, or 90°

10. 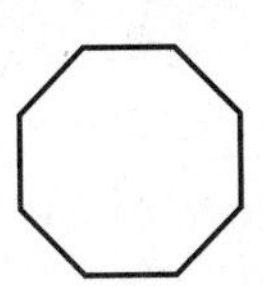

$\frac{1}{8}$ turn, or 45°

11. 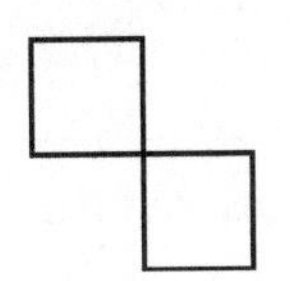

$\frac{1}{2}$ turn, or 180°

12. 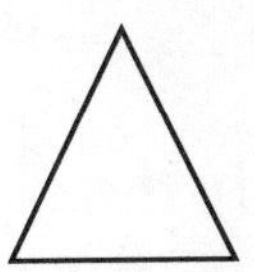

$\frac{1}{3}$ turn, or 120°

13. Print your first and last names in lowercase letters. Identify the letters that have both line symmetry and rotational symmetry. **Possible answers: l, o**
Answers will vary. Check students' work.

Mixed Applications

14. Draw a figure with a vertical line of symmetry. **Answers will vary. Check students' drawings.**

15. The Math Club asked 380 students to name their favorite subject. Math was the favorite subject of 57 students. What percent of the students said math is their favorite subject?

15%

Name ______________________

Transformations

Vocabulary

Write the correct letter from Column 2.

1. image ___d___
2. reflection ___a___
3. rotation ___e___
4. transformation ___b___
5. translation ___c___

a. a figure flipped over a line
b. the movement of a figure
c. a slide of a figure
d. the final position of a figure
e. a figure turned about a point

Describe the transformation that changes the original into the image.

6. FUN → FUN (turned on its side)

90° counter-
clockwise rotation

7. FUN → NUF (flipped)

horizontal
reflection

8. FUN → FUN

translation

Write the word NICE, and use it as the original figure. Then draw the image of each transformation.

9. vertical reflection

NICE → ИICE

10. 180° rotation

NICE → ƎƆIИ

11. 90° clockwise rotation

NICE → NICE (turned on its side)

Identify the type of reflection. Write *vertical, horizontal,* or *diagonal.*

12. 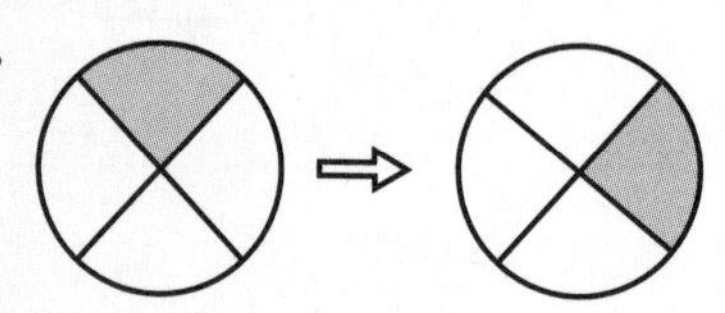

diagonal

13. 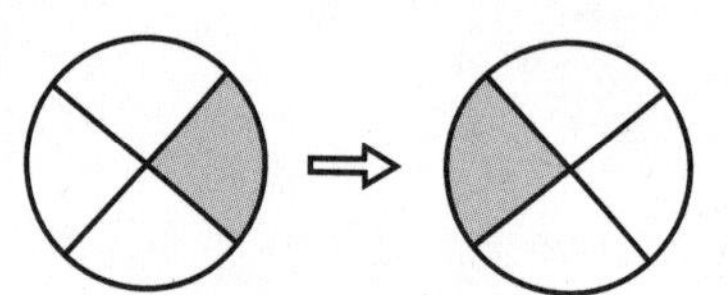

horizontal

14. 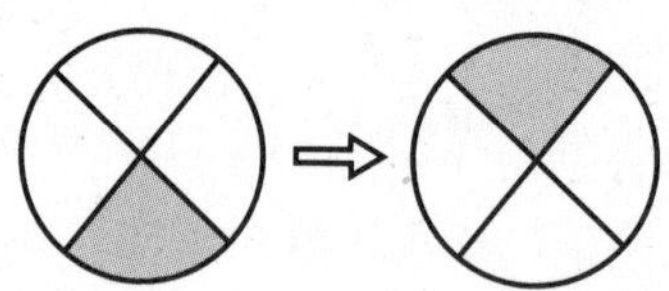

vertical

Mixed Applications

15. Coach Henry posted the batting order upside down on the dugout wall. What type of transformation must he use to right the paper?

rotation

16. Ron bought 12 paintbrushes for $28.68. How much will 25 paintbrushes cost?

$59.75

Use with text pages 200–203.

Name ______________________

LESSON 10.4

Transformations on the Coordinate Plane

Identify the type of transformation. Write *translation, reflection,* or *rotation.*

1. 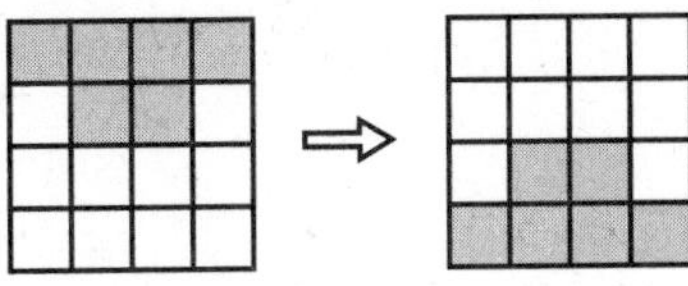

reflection

2. 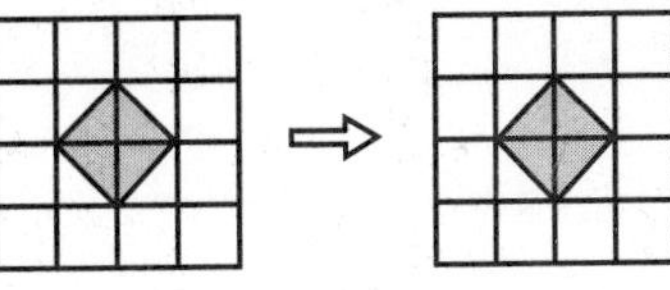

translation

3. 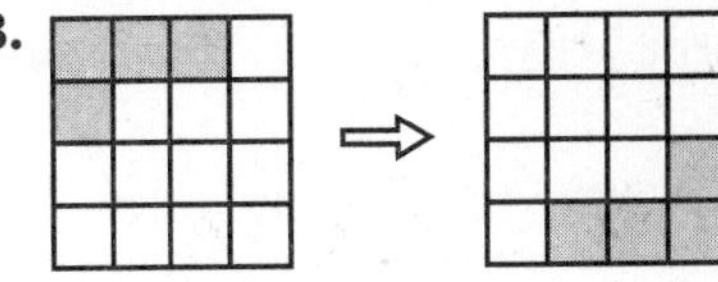

rotation

Begin with figure *ABCDE.* Graph each image and give its coordinates.

4. Reflect about the x-axis.

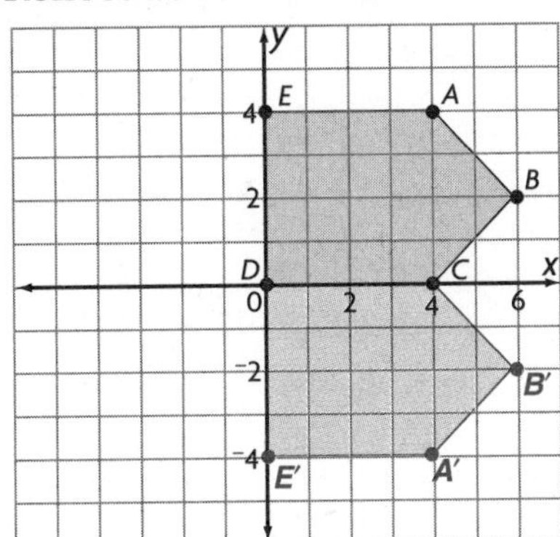

A′ (4,⁻4), B′ (6,⁻2),

C (4,0), D (0,0), E′ (0,⁻4)

5. Rotate 90° clockwise about (0,0).

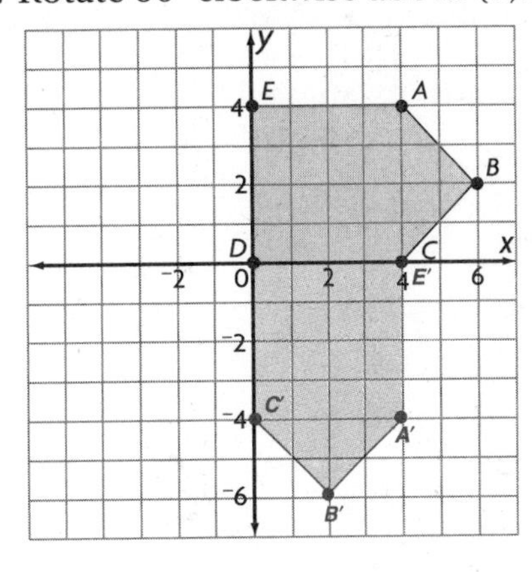

A′ (4,⁻4), B′ (2,⁻6), C′ (0,⁻4),

D (0,0), E′ (4,0)

Mixed Applications

6. Draw a polygon on a coordinate plane. Label the vertices. Translate it 2 units up and 3 units to the right. Then record the new coordinates of the vertices.

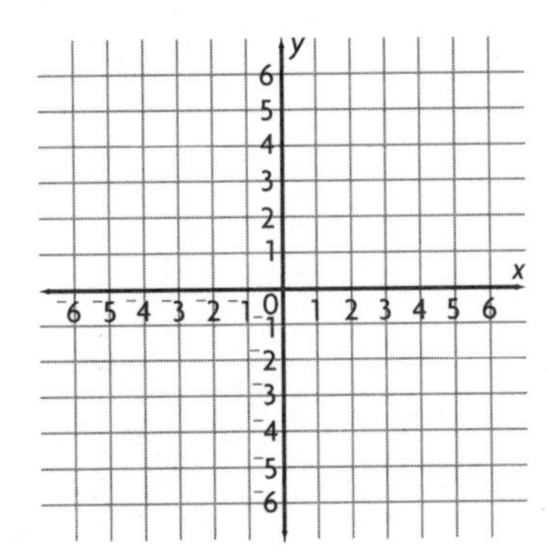

Answers will vary. Check students' work.

7. Draw a trapezoid on a coordinate plane so that one vertex is at (0,0) and another is at (⁻1,⁻2). Rotate the trapezoid 180° about (0,0), and draw the image.

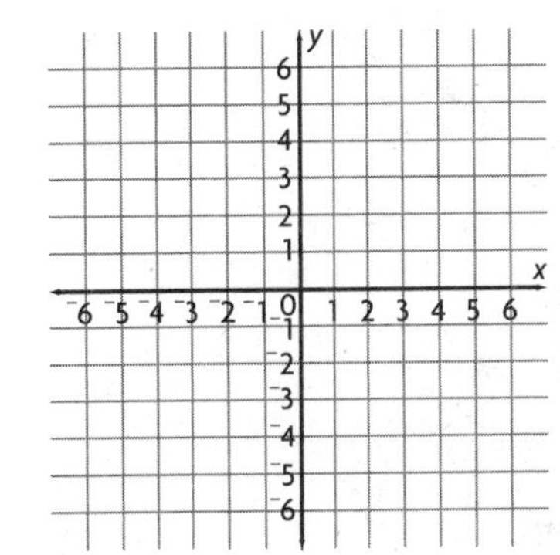

Answers will vary. Check students' work.

Name ______________________

LESSON 11.1

Constructing Congruent Angles and Line Segments

Vocabulary

Complete.

1. The symbol ≅ means **is congruent to**.

2. To **bisect** means to divide into two congruent parts.

3. Constructing a bisector of a line segment is a good way to find a(n) **midpoint** of a segment.

Construct a congruent angle. **Check students' constructions.**

4.

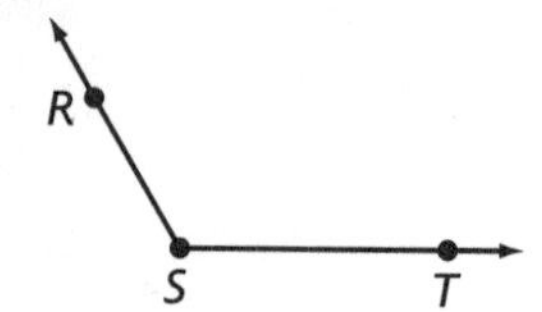

5.

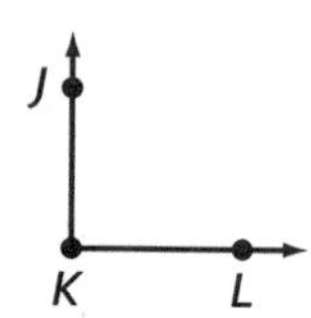

6.

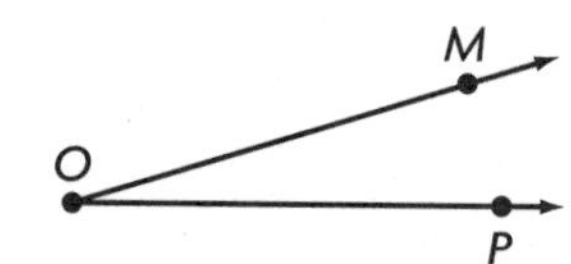

Bisect the figure. **Check students' constructions.**

7.

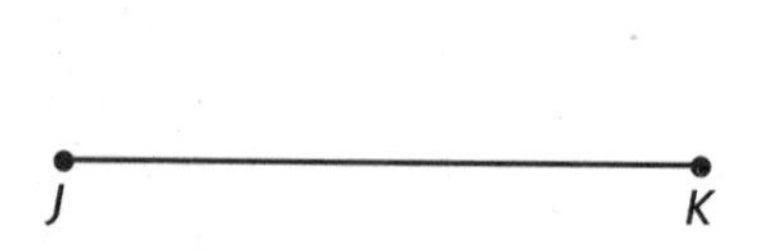

8.

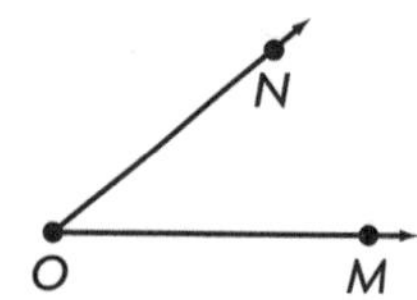

9.

Mixed Applications

10. Use a compass and straightedge to draw a square. Find the midpoint of each side by construction. Connect the midpoints to form a new quadrilateral. What type of quadrilateral is formed?

square

11. Rick is designing a variety of polygons to make a quilt. He can design 6 every $\frac{3}{4}$ hr. How many can he design in $4\frac{1}{2}$ hr?

36 polygons

12. Write a ratio that compares the number of lines of symmetry in an equilateral triangle with the number of sides.

$\frac{3}{3}$ or 1:1

Use with text pages 211–214.

Name ______________________

Constructing Parallel and Perpendicular Lines

Vocabulary

Write the correct letter.

a. 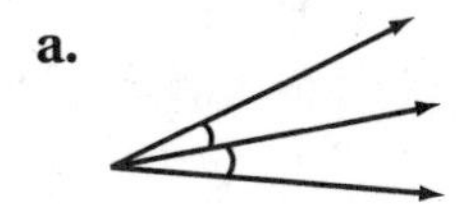

b. P O

1. perpendicular bisector ___b___

---

2. Construct a line through point X parallel to $\overleftrightarrow{YZ}$.
Check students' constructions.

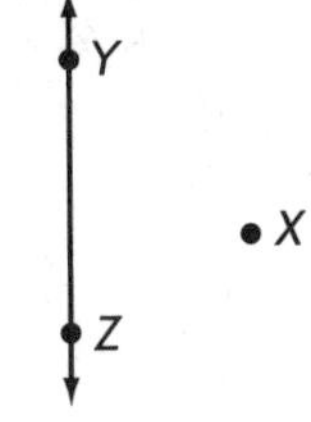

3. Construct a line through point X perpendicular to $\overleftrightarrow{YZ}$.
Check students' constructions.

4. Using a straightedge, draw $\overleftrightarrow{FH}$ in the space below. Using a compass and a straightedge, construct $\overleftrightarrow{GR}$ perpendicular to $\overleftrightarrow{FH}$. Then construct $\overleftrightarrow{LM}$ parallel to $\overleftrightarrow{FH}$.
Check students' constructions.

5. Look at your construction in Exercise 4. What is the relationship between $\overleftrightarrow{LM}$ and $\overleftrightarrow{GR}$? How do you know?

$\overleftrightarrow{LM}$ is perpendicular to $\overleftrightarrow{GR}$.

Possible explanation: The angle where $\overleftrightarrow{LM}$ and $\overleftrightarrow{GR}$ intersect measures 90°.

Mixed Applications

6. Name three examples of parallel lines found in your classroom, community, or home.

Answers will vary.

7. Name three examples of parallel lines intersecting perpendicular lines found in your classroom, community, or home.

Answers will vary.

8. Of the 28 pictures that Rose painted, $\frac{3}{4}$ contain perpendicular lines. How many pictures do not contain perpendicular lines?

7 pictures

9. Suppose you bisect a 100° angle. Next, you bisect one of the smaller angles formed. You then bisect one of those angles. What is the measure of the smallest angle formed?

12.5°

Name ______________________

LESSON 11.3

Classifying and Comparing Triangles

Vocabulary

Write *ASA* for Angle-Side-Angle, *SAS* for Side-Side-Side, or *SSS* for Side-Side-Side.

1. Two sides and the included angle of one triangle are congruent to two sides and the included angle of another triangle.

SAS

2. Three sides of one triangle are congruent to three sides of another triangle.

SSS

3. Two angles and the included side of one triangle are congruent to two angles and the included side of another triangle.

ASA

Classify each triangle according to the lengths of its sides.

4. 15 cm, 15 cm, 15 cm

equilateral

5. 35 m, 25 m, 15 m

scalene

6. 9 in., 12 in., 9 in.

isosceles

Classify each triangle according to the measures of its angles.

7. 10°, 25°, 145°

obtuse

8. 90°, 20°, 70°

right

9. 40°, 80°, 60°

acute

Determine whether the triangles are congruent by SSS, SAS, or ASA.

10.

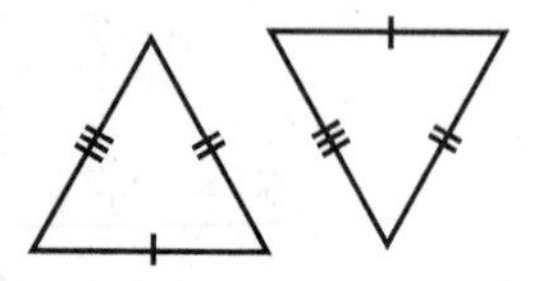

SSS

11.

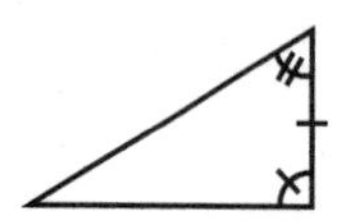

ASA

12.

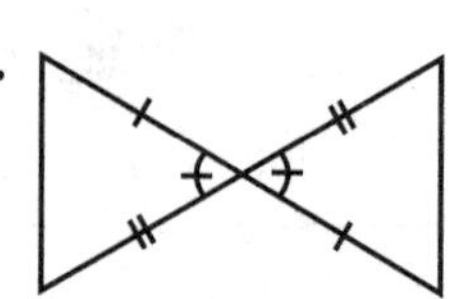

SAS

Mixed Applications

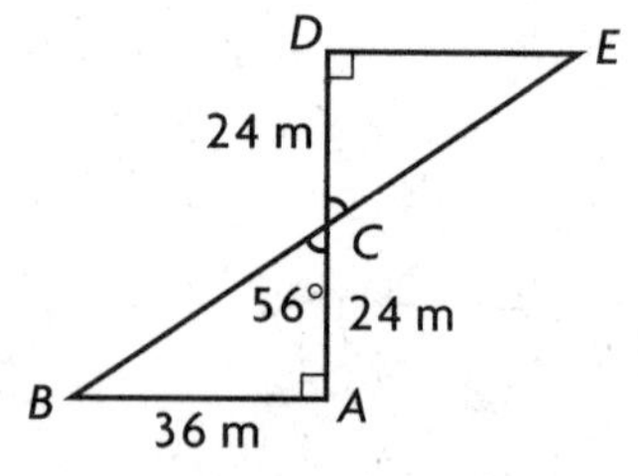

13. Find the measure of $\angle CED$. Give a reason for your answer.

34°; $\triangle ACB \cong \triangle DCE$ by ASA, so

$\angle ABC \cong \angle DEC$, measure of

$\angle ABC = 180 - 90 - 56 = 34°$

14. Construct a segment perpendicular to $\overleftrightarrow{MN}$. Check students' work.

Use with text pages 218–221.

Name ____

Constructing Congruent Triangles

Use the indicated rule to construct a congruent triangle. **Check students' constructions.**

1. SAS

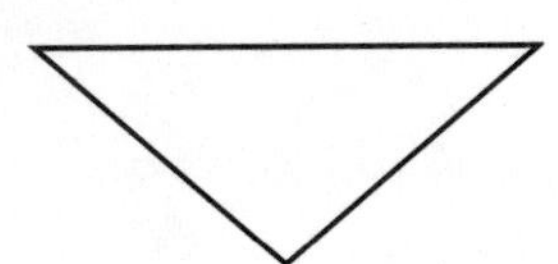

2. ASA

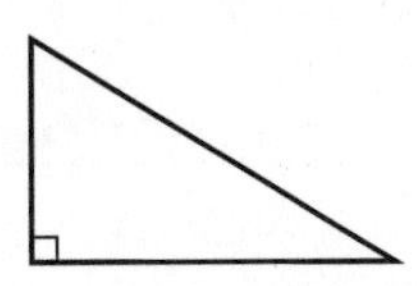

3. SSS

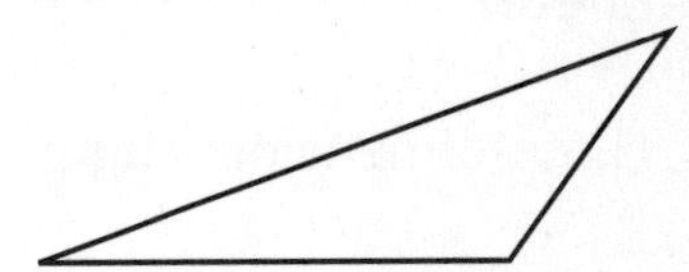

Use the SSS rule and the three given segments to try to construct a triangle. **Check students' constructions.**

4.

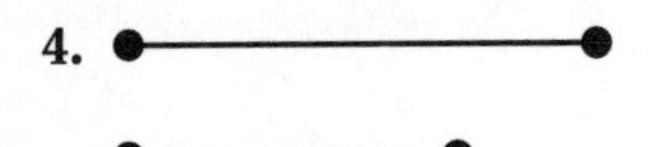

5.

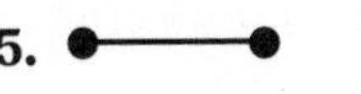

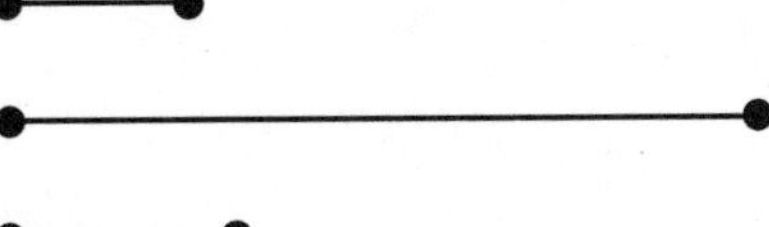

6.

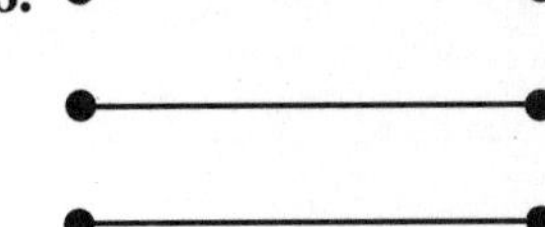

7. In which of Exercises 4–6 was the construction not possible? Explain.

Exercise 5; the three segments do not form a triangle.

Mixed Applications

8. Can you use the SSS rule to construct a triangle whose sides measure $2\frac{1}{2}$ in., $\frac{1}{2}$ in., and $\frac{1}{2}$ in.? Explain.

No; $\frac{1}{2} + \frac{1}{2} < 2\frac{1}{2}$

9. Explain how the rules for congruence of triangles help you construct congruent triangles.

The rules tell what corresponding parts to use for construction.

In Exercises 10 and 11, identify the transformation that forms the second figure.

10.

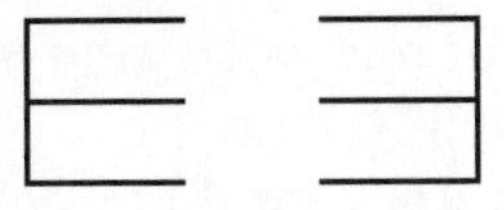

reflection

11.

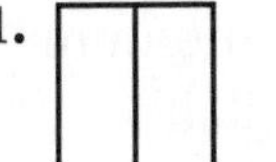

rotation

Use with text pages 222–225.

Name ______________________________

LESSON 12.1

Solid Figures

Vocabulary

Complete.

1. Three-dimensional figures are called **space figures** or solid figures.

2. If all the faces of a solid figure are polygons, the figure is a(n) **polyhedron**.

Write *polyhedron* or *not polyhedron* for each solid figure.

3. 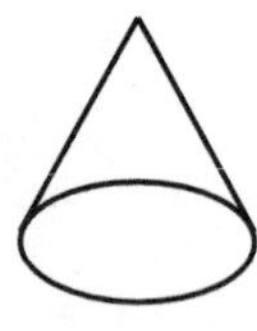

not polyhedron

4. 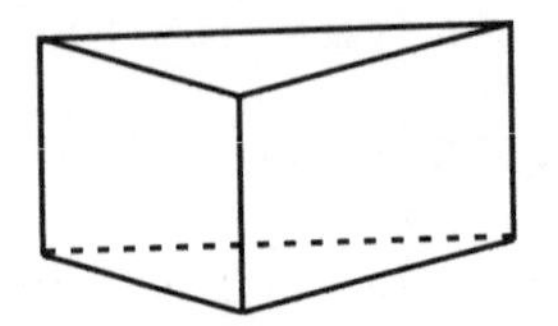

polyhedron

5. 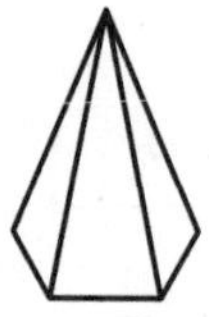

polyhedron

Name the figure that is the base of each solid figure.

6. 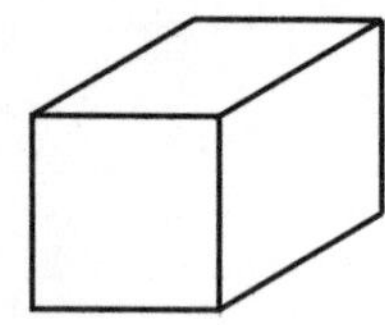

rectangle

7. 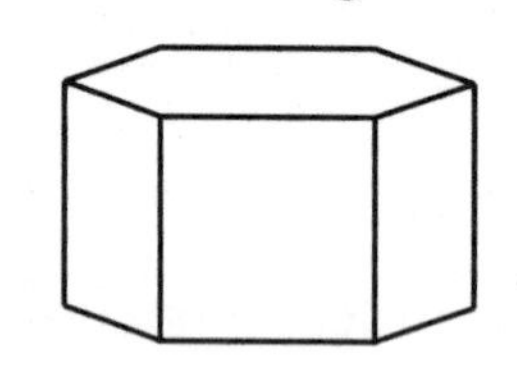

hexagon

8. 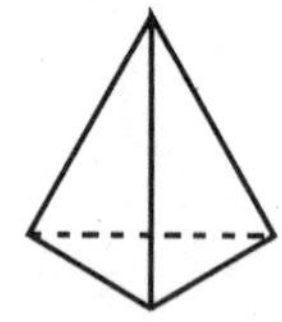

triangle

Write *true* or *false.* Write a true statement from any false statement.

9. Some faces of a pyramid are triangles.

true

10. All cubes are polyhedrons.

true

11. Every cone is a polyhedron.

False; no cone is a polyhedron.

12. All polyhedrons are prisms.

False; some polyhedrons are not prisms.

Possible statements are given for Exercises 11 and 12.

Mixed Applications

13. Alieu says that the globe in his homeroom is a polyhedron. Is he correct? Explain.

No; a globe is a sphere, with a curved surface.

14. WXY's stock prices for one week were $35\frac{1}{4}$, $36\frac{1}{2}$, $35\frac{7}{8}$, 37, and $36\frac{5}{8}$. Find the average price of the stock for that week.

$36\frac{1}{4}$

Use with text pages 230–231.

Name ______________________

Problem-Solving Strategy

Finding Patterns in Polyhedrons

For Problems 1–2, use a pattern. For Problem 3, find a pattern.

1. A hexagonal prism has 8 faces and 12 vertices. How many edges does it have?

   $8 + 12 - 2 = 18$; 18 edges

2. Dwight is building a rectangular pyramid. He knows it will have 5 faces and 8 edges. How many vertices will it have?

   $8 = 5 + V - 2$; $V = 5$; 5 vertices

3. As Amy is building prisms, she notices that the triangular prism has 6 vertices, the rectangular prism has 8 vertices, and the pentagonal prism has 10 vertices. How many vertices will she find on the octagonal prism? Sketch an octagonal prism to check your prediction.

   Check students' sketches; 16 vertices.

Mixed Applications

Solve.

CHOOSE A STRATEGY

- Find a Pattern
- Use a Formula
- Guess and Check
- Draw a Diagram
- Write an Equation
- Make a Table

Choices of strategies will vary.

4. Rod is building a playhouse in the shape of a prism that has 12 faces. He needs special angle irons at each vertex. How many vertices are in Rod's playhouse?

   20 vertices

5. During a sale, The Music Shoppe sells twice as many CDs as tapes. A total of 51 tapes and CDs are sold. How many of each are sold?

   17 tapes and 34 CDs

6. Use the signs + and − to complete the expression so that it has a value of $^-2$.

   11 [+] 4 [−] 10 [−] 7

7. Ali drove for $2\frac{1}{2}$ hr. Her average speed was 48 mi per hour. How far did she drive?

   120 mi

8. Roger is 5 years older than his sister, who is $\frac{2}{3}$ the age of their brother Sam. Sam is 12 years old. How old is Roger?

   13 years old

9. The temperature in Westboro one cool summer day was 20° C. What was the temperature in °F? Use the formula $F = \frac{9}{5}C + 32$.

   68° F

Name ______________________________

Nets for Solid Figures

Vocabulary

Complete.

1. An arrangement of polygons in a plane is called a(n) ___net___ if the arrangement can be folded to form a polyhedron.

Name the prism or pyramid that can be formed from the net.

2. 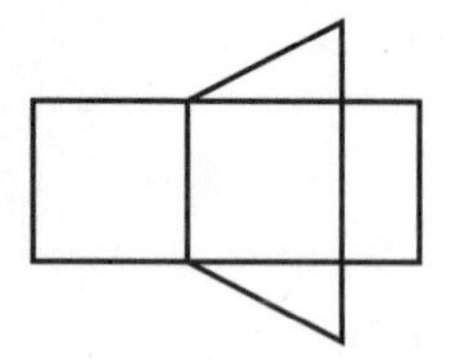

___triangular prism___

3. 

___pentagonal pyramid___

4. 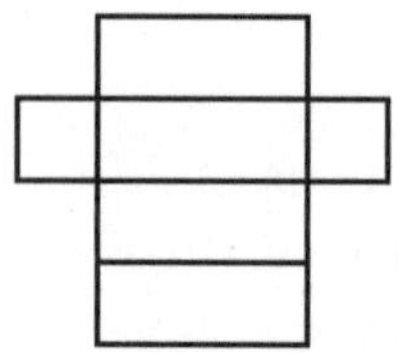

___rectangular prism___

Tell if each arrangement of squares will make a cube.

5. 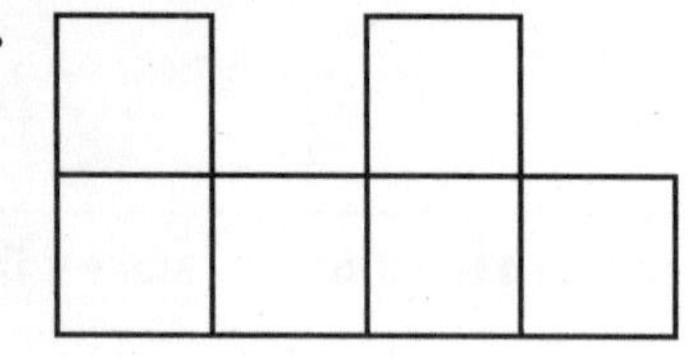

___no___

6. 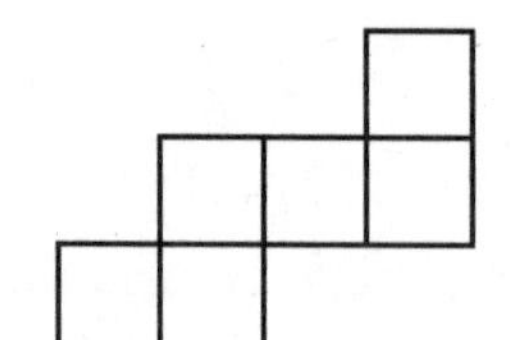

___yes___

7. 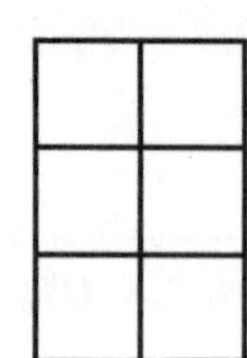

___no___

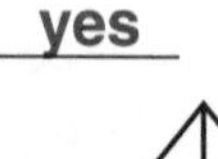

For Exercises 8–9, look at the square pyramid at the right.

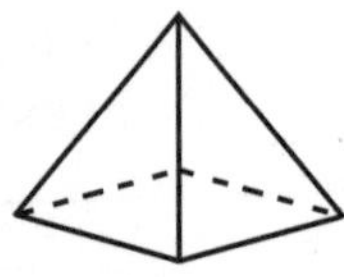

8. How many faces does the pyramid have? What types of polygons are the faces?

___5 faces; 4 triangles and 1 square___

9. Draw a net for the prism.

Check students' nets.

Mixed Applications

10. Visualize the net at the right folded into a cube. Which face will be on top if Face A is on the bottom?

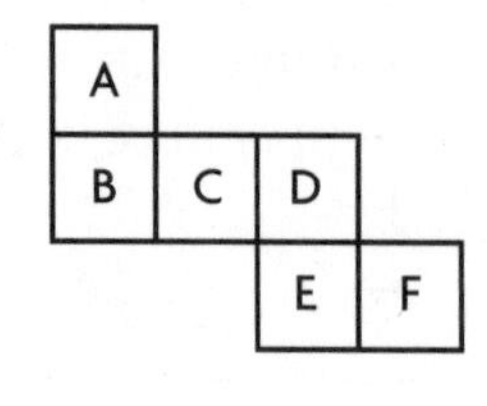

___Face E___

11. Buzz saves 15% of his weekly salary. He earns $79.57. About how much does he save, to the nearest dollar?

___about $12___

Use with text pages 237–239.

Name ______________________

LESSON 12.4

Drawing Three-Dimensional Figures

Vocabulary

Complete.

1. The faces of a polyhedron that are not bases are called lateral faces.

For Exercises 2–3, look at the pentagonal prism at the right.

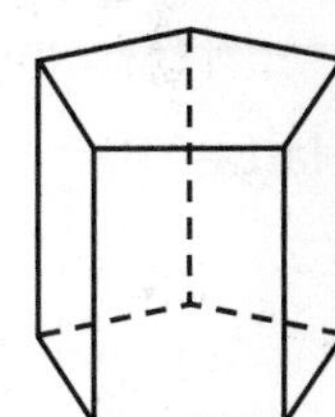

2. How many faces are visible? 4 faces

3. How many edges and vertices are visible? 12 edges, 9 vertices

For Exercises 4–6, use the figures shown below.

4. Which figure(s) has a circle in one of its views? d

5. Which figure(s) has a triangle in one of its views? a, b

6. Which figure(s) has a rectangle in one of its views? a, c

a. 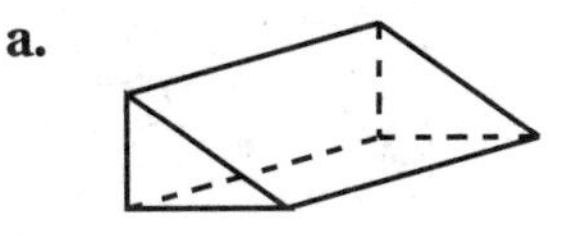

b. 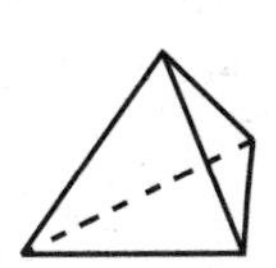

c. 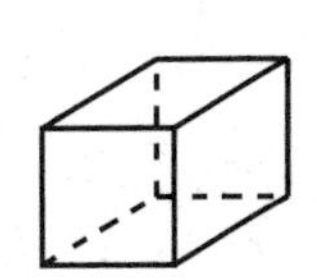

d. 

Draw and name the figure you would see in each view of the solid figure. Check students' drawings.

7. Top View

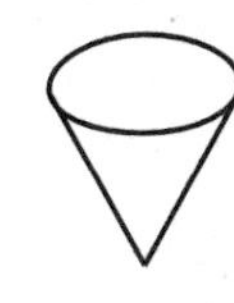

circle

8. Bottom View

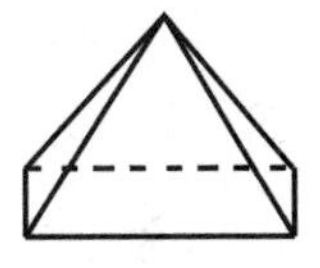

rectangle

9. Side View

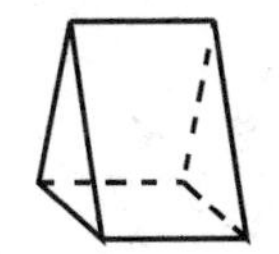

triangle

10. Back View

rectangle

Use dashed lines to show the hidden edges.

11. Triangular Pyramid

12. Hexagonal Prism

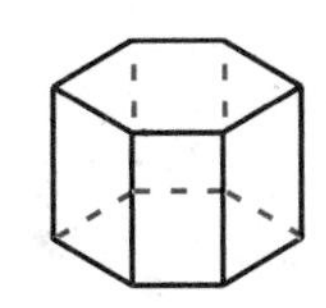

13. Square Pyramid

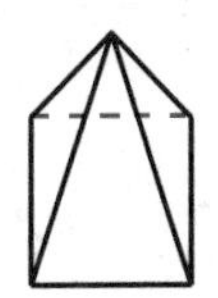

14. Triangular Prism

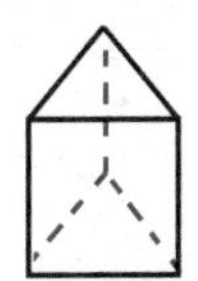

Mixed Applications

15. Rick is drawing a solid figure that has 4 sides. All the sides are triangles. What solid figure is he drawing?

triangular pyramid

16. How many pieces of string 3.5 ft long can Glenda cut from 49 ft of string.

14 pieces

Name ___________________________

Tessellations

Vocabulary

Complete.

1. A tessellation is a(n) **repeating** pattern of **congruent** plane figures.

2. A tessellation completely covers a(n) **plane** with no gaps or **overlaps**.

3. A figure that is repeated to make a pattern is a(n) **basic unit**.

Use the figure(s) to try to make at least two rows of a tessellation. Write *yes* or *no* to tell whether a tessellation can be made. **Check students' tessellations.**

4. scalene triangle **yes**

5. regular octagon and square **yes**

6. regular hexagon **yes**

7. regular octagon **no**

Cut the basic unit out of graph paper. Use it to create a tessellation of at least two rows. **Check students' tessellations.**

8. 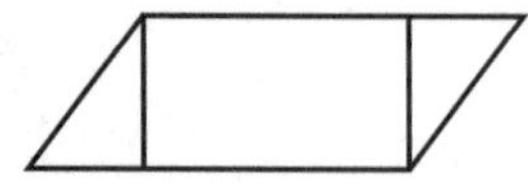

9. 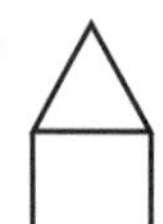

10. 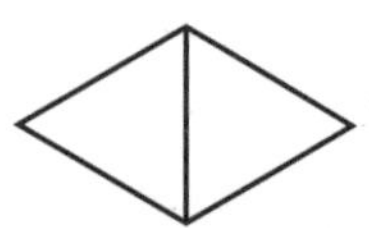

Mixed Applications

11. Draw a rectangle. Use translations to change the shape of the rectangle. Use the figure to create at least two rows of a tessellation. **Check students' tessellations.**

12. Name the prism that can be formed from the net. How many faces does it have? How many vertices? How many edges?

triangular prism; 5 faces; 6 vertices; 9 edges

13. Renting a car costs $31 per day plus $0.25 per mile. How far can Meryl travel in 1 day if she has only $76? in 2 days?

180 mi; 56 mi

14. Draw three different isosceles triangles. Try to create a tessellation with each figure. Do you think all isosceles triangles can form tesselations?

Check students' tessellations; yes.

Use with text pages 250–252.

Name ______________________

LESSON 13.2

Geometric Iterations

For Exercises 1–5, rotate the circle 90° clockwise. Complete the iteration process five times. Draw the figure at each stage.

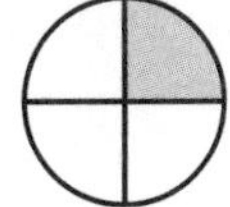

Check students' drawings.

1. Stage 1

2. Stage 2

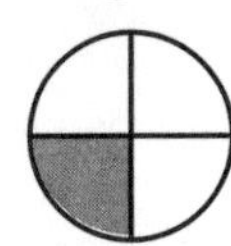

3. Stage 3

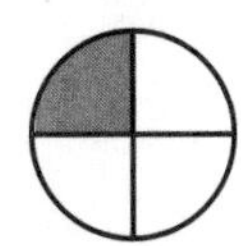

4. Stage 4

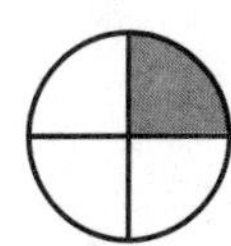

5. Stage 5

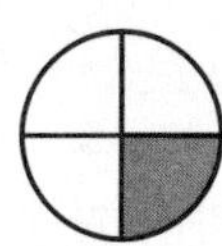

6. When do the positions repeat themselves? at every fourth stage

Using the figure above, rotate the circle 180° clockwise. Complete the iteration process five times. Draw the figure at each stage. Check students' drawings.

7. Stage 1

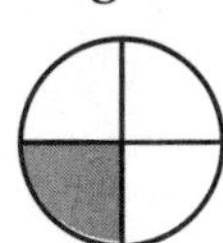

8. Stage 2

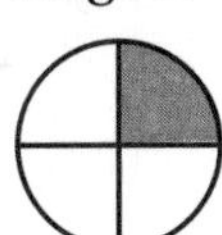

9. Stage 3

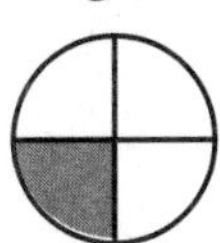

10. Stage 4

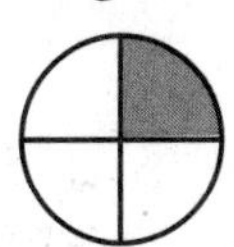

11. Stage 5

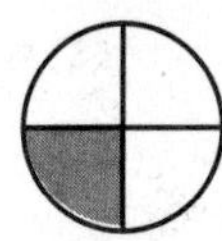

12. When do the positions repeat themselves? at every even-numbered stage

Trace the hexagon and rotate it 60° clockwise. Complete the iteration process three times. Write the number where red stops and the number where blue stops.

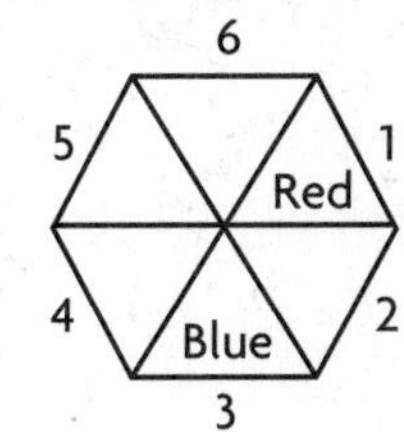

13. Stage 1 red: 2; blue: 4

14. Stage 2 red: 3; blue: 5

15. Stage 3 red: 4; blue: 6

Name the part of the hexagon where red stops at the given stage of iteration.

16. Stage 9 4

17. Stage 6 1

18. Stage 12 1

Name the part of the hexagon where blue stops at the given stage of iteration.

19. Stage 5 2

20. Stage 4 1

21. Stage 6 3

Mixed Applications

22. Draw and label a segment of length 3 cm. Rotate the segment 30° clockwise. How many times do you have to repeat the iteration process to reach the original segment?

12 times

23. The senior class of Cambridge Central School sold $\frac{3}{4}$ of the tickets to the senior prom. If 75 tickets were not sold, how many tickets did the class sell?

225 tickets

Name ____________________

Self-Similarity

Vocabulary

Complete.

1. Similar figures are figures with the same **shape** but not necessarily the same **size**.

2. A figure has self-similarity if **smaller parts look like smaller copies of the whole figure**.

Each of these drawings shows a stage in an iteration process. In each answer for Exercises 3–6, refer to all three drawings.

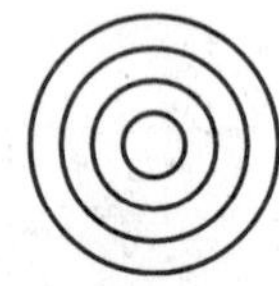
Figure A

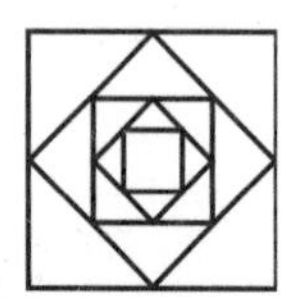
Figure B

Figure C

3. What geometric figure or figures are used? ***A*: circle; *B*: square; *C*: circle, regular pentagon, equilateral triangle**

4. Do the figures you described in Exercise 3 have self-similarity? Explain.

   ***A*: yes, same shape; *B*: yes, same shape; *C*: no, different shapes**

5. What new figure or figures would appear at the next stage?

   ***A*: circle; *B*: square; *C*: answers will vary.**

6. Would the figures at the next stage have self-similarity? Explain.

   ***A*: yes, same shape; *B*: yes, same shape; *C*: no, different shape**

Mixed Applications

7. Does the figure have self-similarity?

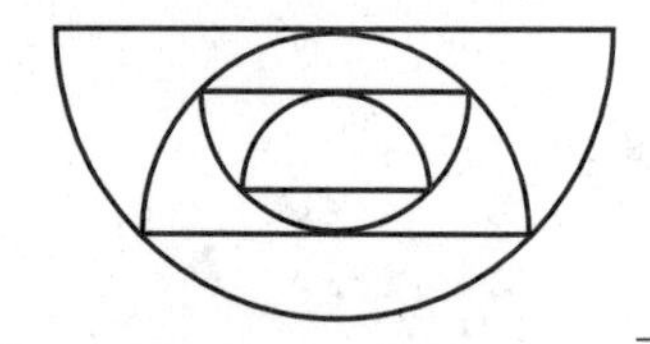

yes

8. The attendance today at Miller Middle School is 93% of the school's enrollment. The attendance is 372 students. What is the school's enrollment?

   400 students

Use with text pages 255–257.

Name ______________________________

LESSON 13.4

Fractals

Vocabulary

Complete.

1. A ___fractal___ is a repeating pattern containing shapes that are alike but of different sizes throughout.

For Exercises 2–3, use the figures shown.

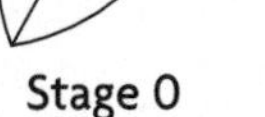

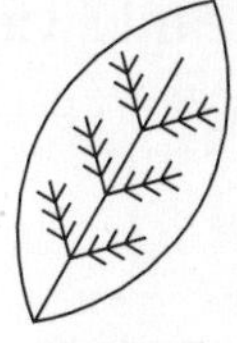

Stage 0　　Stage 1　　Stage 2

2. How many new branches do you see at Stage 1?

   at Stage 2? ___6 branches; 36 branches___

3. How many new branches would there be at Stage 3?

   at Stage 4? at Stage *n*? ___216 branches; 1,296 branches; 6^n branches___

For Exercises 4–7, use the figures below.

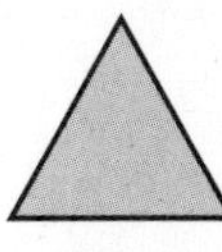
Stage 0

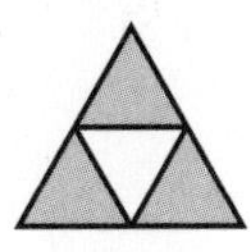
Stage 1

Stage 2

Stage 3

4. How does the length of a side of the equilateral triangle change from stage to stage? ___decreases by $\frac{1}{2}$___

5. How many copies of the Stage 0 figure are used to build the Stage 1 figure? ___3 copies___

6. How is Stage 1 built with the copies of the Stage 0 figure?

   ___by placing a copy along each side___

7. In the space above, draw Stage 3 of the iteration process. How many shaded triangles do you see at Stage 3? ___27 triangles___ Check students' drawings.

Mixed Applications

8. Draw an isosceles triangle. Rotate the triangle 45° counterclockwise about one vertex. How many times do you have to complete the iteration process to reach the original triangle?

   ___8 times___

9. Ian sold 8 fewer tickets to the school play than Dan. Together the boys sold 32 tickets. How many did each boy sell?

   ___Dan, 20 tickets; Ian, 12 tickets___

Name ______________________________

LESSON 14.1

Problem-Solving Strategy

Drawing a Diagram to Show Ratios

Draw a diagram and solve. **Check students' diagrams.**

1. Matt mixes custom paints. He mixes red paint and white paint in a ratio of 4 to 7 to create a special shade of rose. How many gallons of red paint should he mix with 14 gal of white to make the rose paint?

   8 gal

2. Rachael's pet cricket jumps farther than Diego's pet grasshopper. The ratio of the length of the cricket's jump to the length of the grasshopper's jump is 4 to 3. If the insects start at the same time and place and they race a total of 24 meters, where will they land at the same places?

   at 12 m and at 25 m

3. Ferry Avenue Florist sells wedding arrangements that use roses and carnations in a ratio of 2 to 3. Sandy's sister's wedding arrangements will use 147 carnations. How many roses will the florist need?

   98 roses

4. A path is made of different sizes of brick. One row uses bricks 4 in. wide, and another row uses bricks 6 in. wide. In how many inches will the rows be even and the pattern of bricks start to repeat?

   12 in.

Mixed Applications

Solve.

CHOOSE A STRATEGY

- Make a Table
- Look for a Pattern
- Draw a Diagram
- Make a List

Choices of strategies will vary.

5. Richard has twice as many marbles as Kathy. Matthew, who has 9 marbles, has 3 fewer than Kathy. How many marbles does Richard have?

   24 marbles

6. During the past five weeks, Lara has saved \$6, \$8, \$12, \$18, and \$26. If the pattern continues, how much will Lara save in the seventh week?

   \$48

7. Anna needs ribbon for the costume she will wear in the school play. She needs $11\frac{3}{8}$ in. for the neck and $7\frac{1}{6}$ in. for each sleeve. How many inches of ribbon will she have left if she buys 1 yd of ribbon?

   $10\frac{7}{24}$ in.

8. It costs \$0.32 to mail a letter and \$0.20 to mail a postcard. Walter wrote to 20 people in December. He paid \$5.68 for postage during the month. How many postcards did he write in December?

   6 postcards

9. The ratio of boys to girls in Liam's gym class is 5 to 3. How many boys are in the class if there are 27 girls?

   45 boys

10. When Larry opens a book, the product of the numbers of the two facing pages is 7,656. What are the page numbers?

    pages 87 and 88

Use with text pages 270–271.

Name ____________________

Ratios and Rates

Vocabulary

Write *true* or *false*.

1. A *rate* compares two quantities of the same kind. **false**

2. A *unit rate* is a ratio in which the denominator is 1. **true**

3. Comparing *unit prices* is a useful way to compare values. **true**

Use the given rate of pay and the time worked to find how much is earned.

4. \$3.75 per hour for 28 hr **\$105**

5. \$7 per hour for $8\frac{1}{2}$ hr **\$59.50**

Find the unit rate.

6. 8 for \$26, or **\$3.25** each

7. 252 km in 6 hr, or **42** km per hr

Find each unit price. Round to the next cent when necessary. Then tell which choice has the lower unit price.

8. a 15-oz bottle of shampoo for \$1.63 or a 30-oz bottle for \$2.99

   \$0.11; \$0.10; 30-oz bottle

9. a 5-lb bag of oranges for \$4.25 or a 12-lb bag for \$10.20

   \$0.85; \$0.85; unit prices are equal.

10. a package of 6 eggs for \$0.78 or a dozen eggs for \$1.31

    \$0.13; \$0.11; a dozen eggs

11. 2 yd of fabric for \$6.98 or 3 yd for \$9.87

    \$3.49; \$3.29; 3 yd of fabric

Mixed Applications

12. In a gourmet mix, the ratio of white jelly beans to green jelly beans is 2 to 5. If a bag contains 24 white jelly beans, how many are green?

    60 green jelly beans

13. Mrs. Schwede calls her mother every week. A 15-min long-distance phone call costs \$2.25. What is a reasonable estimate of the cost of a 40-min call?

    Possible answer: \$6.00

14. Majestic Theater sells a 16-oz bag of popcorn for \$2.50 and a 36-oz box for \$4.50. Which is the better buy?

    the 36-oz box

15. The width of a mirror is 2 in. less than its length. The area of the mirror is 168 in.^{2} Find the dimensions of the mirror.

    12 in. × 14 in.

Use with text pages 272–275.

Name ______________________________

LESSON 14.3

Rates in Tables and Graphs

1. The table shows the cost of photocopying at Oliver's Office On-line. If the cost per page does not change, how much does it cost to photocopy 550 pages? **$30.25**

PHOTOCOPYING AT OLIVER'S OFFICE ON-LINE	
Number of Pages	Cost
100	$5.50
150	$8.25
200	$11.00

2. Whitney sold wrapping paper as a fund-raiser for her school band. She kept a record of her sales and profits in a table. Look for patterns in her data. Then use the patterns to complete the table.

Packages Sold	5	10	15	20
Money Collected	$25.55	$51.10	**$76.65**	**$102.20**
Profit for Band	$10.55	$21.10	**$31.65**	**$42.20**

3. The table shows the values of Mr. Vincent's $30,000 sports car one, two, and three years after he buys it. Compare the values for any two years by making a ratio. Use the ratio to find a pattern in the table. Describe the pattern.

Possible answer: $24,000 to $30,000 is 0.8 to 1;

pattern: multiply by 0.8.

Year	0	1	2	3
Value	$30,000	$24,000	$19,200	$15,360

4. If the pattern continues, what will be the value of Mr. Vincent's sports car four years after he buys it? five years after he buys it? **$12,288; $9,830.40**

Mixed Applications

5. Use the graph at the right. What does the post office charge for the first ounce? for each ounce after the first? How much does the post office charge for a 3-oz letter?

$0.32; $0.23; $0.78

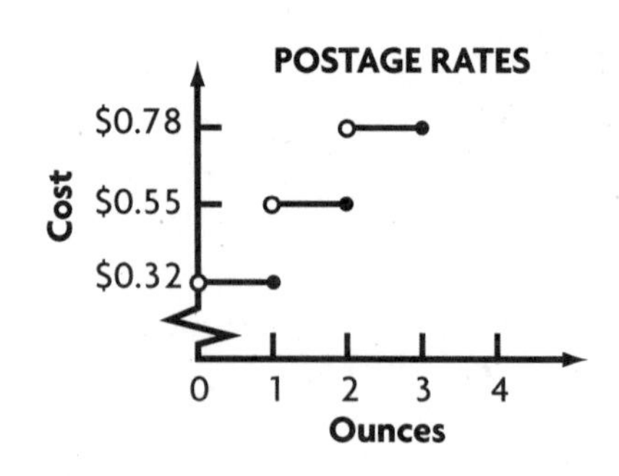

6. Mr. Allen is an amateur photographer who specializes in underwater photography. He can buy 6 rolls of film for $19.75 or 9 rolls of film for $29.25. Which choice has the lower unit price?

9 rolls for $29.25

Use with text pages 276–277.

Name ____________________

Finding Golden Ratios

Vocabulary

Complete.

1. A ratio that is approximately equal to 1.61 is called a(n) **Golden Ratio**.
2. The division of a segment into two parts that form a Golden Ratio is called a(n) **Golden Cut**, or a(n) **Golden Section**.

For the given segment, mark a point, *C*, that makes a Golden Cut.

3. **C should be marked at about 6.9**

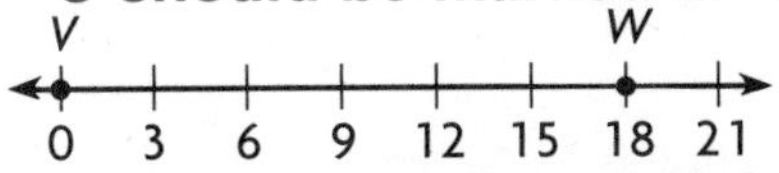

or at about 11.1.

4. **C should be marked at about 15.2**

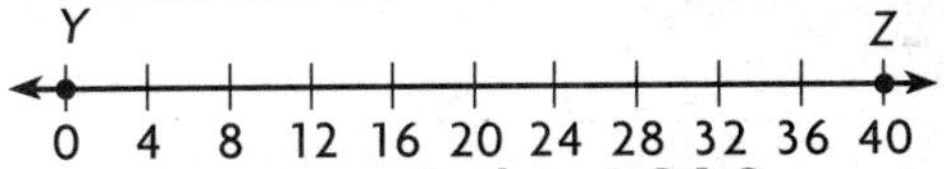

or at about 24.8.

5. Point *R* divides $\overline{ST}$ so that $\overline{RS}$ = 30 m and $\overline{RT}$ = 16 m. Does point *R* come close to making a Golden Cut? Support your answer.

   no; $\frac{30}{16} \approx 1.88 \neq 1.61$

6. Segment *AC* is 52 in. long. Where should point *B* be placed on $\overline{AC}$ to make $\frac{AC}{BC}$ a Golden Ratio? **about 32.3 in. from *A***

Mixed Applications

7. Helen is designing a shed that will measure 26 ft in width. She wants the ratio of length to width to be the Golden Ratio. What should be the length of the shed, rounded to the nearest foot?

   42 ft

8. During the past four years, Jill's bank account earned $36, $42, $54, and $72 in interest. If the pattern continues, how much interest will Jill earn in the eighth year?

   $204

9. The circus is coming to town, and Gyro the giraffe has her picture in the newspaper. In the picture, Gyro has the following dimensions: neck, 4.8 cm; leg, 3 cm; total height, 8.9 cm. Which of Gyro's dimensions approximate the Golden Ratio?

   $\frac{\text{neck}}{\text{leg}} = \frac{4.8 \text{ cm}}{3 \text{ cm}} = 1.6$

10. Tommy is helping his cousin mix paint. To make green paint, they must mix blue and yellow paint in a ratio of 3 to 2. How many quarts of blue paint should they mix with 18 qt of yellow to make green paint?

    27 qt

Name ______________________

LESSON 15.1

Changing Ratios to Percents

Vocabulary

1. What does the word *percent* mean? per hundred

2. You can change a ratio to a percent by using equivalent ratios or a proportion.

Write each ratio as a percent.

3. 3:25 12%

4. 30 to 40 75%

5. $\frac{7}{10}$ 70%

6. 15 to 30 50%

7. 16:1,000 1.6%

8. $\frac{11}{20}$ 55%

9. $\frac{5}{2}$ 250%

10. 4:200 2%

11. 6:5 120%

Use a proportion to write each ratio as a percent.

12. 1 to 3 $33\frac{1}{3}\%$

13. 12:300 4%

14. $\frac{3}{8}$ $37\frac{1}{2}\%$

Complete the table.

	Ratio	Percent
15.	1 to 8	12.5%
16.	14 to 10	140%
17.	$\frac{2}{5}$	40%

	Ratio	Percent
18.	3 to 10	30%
19.	3 to 2	150%
20.	7 to 8	87.5%

Mixed Applications

21. In the basketball playoffs, Susan made 6 of 15 free throws and Alexis made 12 of 16. Who made the greater percent of free throws? Explain.

Alexis; 75% > 40%

22. A diner serves hamburgers that use $\frac{1}{2}$ lb ground beef. How much ground beef is needed to make 18 hamburgers?

9 lb

23. A basketball court is shaped like a rectangle. The length is 88 ft, and the width is 52 ft. What is the area of the court?

4,576 ft^2

24. A box contains 7 red balls, 3 green balls, and 5 yellow balls. What percent of the balls are yellow?

$33\frac{1}{3}\%$

Name ______________________

Finding a Percent of a Number

Vocabulary

Complete.

1. In an equation based on a sentence, the word *is* is represented by a(n) equals sign, *of* by a(n) multiplication sign, and *what number* by a(n) variable.

Find the percent of each number.

2. 10% of 75 7.5
3. 45% of 89 40.05
4. 12% of 180 21.6
5. 0.3% of 16 0.048
6. 40% of 320 128
7. 125% of 420 525
8. 250% of 38 95
9. 67% of 500 335
10. 1% of 59 0.59
11. 100% of 84 84
12. 7% of 79.3 5.551
13. 25% of 117 29.25
14. 82% of 54 44.28
15. 0.5% of 25 0.125
16. 50% of 5 2.5

Use mental math to find the percent of each number.

17. 45% of 20 9
18. 38% of 100 38
19. 80% of 40 32
20. 125% of 400 500
21. 24% of 50 12
22. 30% of 10 3

Mixed Applications

23. In Pennsylvania a sales tax of 6% is charged on nonfood and nonclothing purchases. What is the sales tax on a purchase of $98 for sports equipment?

$5.88

24. Beth bought a coat for 30% off the original price. The original price was $75. What was the sale price of the coat?

$52.50

25. One box of nails weighs 7.6 lb, and another box weighs 12.16 lb. How much more does the heavier box weigh? What is the ratio of the heavier box to the lighter box?

4.56 lb; 1.6 to 1, or $\frac{8}{5}$

26. By the end of the day, $\frac{3}{4}$ of Minerva's room has been painted. Max did $\frac{2}{3}$ of the painting. How much of the room did Max paint?

Max painted $\frac{1}{2}$ of the room.

Name ______________________

Finding What Percent One Number Is of Another

Draw a diagram. Then solve. **Check students' diagrams.**

1. What percent of 40 is 8? **20%**

2. What percent of 24 is 18? **75%**

Write an equation. Then solve.

3. What percent of 10 is 2? $p \times 10 = 2; p = 0.20$, or 20%

4. What percent of 64 is 16? $p \times 64 = 16; p = 0.25$, or 25%

Write a proportion. Then solve.

5. What percent of 50 is 54? $\frac{54}{50} = \frac{n}{100}$; $n = 108$; 108%

6. What percent of 80 is 140? $\frac{140}{80} = \frac{n}{100}$; $n = 175$; 175%

Solve. Use any of the methods you have studied.

7. What percent of 12 is 3? **25%**

8. What percent of 36 is 9? **25%**

9. What percent of 88 is 11? **about 12%**

10. 29 is what percent of 116? **25%**

11. What percent of 20 is 8? **40%**

12. What percent of 35 is 35? **100%**

13. What percent of 50 is 25? **50%**

14. What percent of 90 is 81? **90%**

15. 13 is what percent of 15? **about 87%**

16. What percent of 25 is 50? **200%**

Mixed Applications

17. In Amy's math class, 40% of the students finished the math assignment before lunch. There are 35 students in the class. How many students finished before lunch?

14 students

18. To earn a B in his science class, Ted must score at least 80% on his test. He answered 12 of 16 questions correctly. Did Ted score high enough to get a B? Explain.

No; 75% < 80%.

Use with text pages 293–295.

Name ____________________

LESSON 15.4

Finding a Number When the Percent Is Known

Solve.

1. 25% of what number is 16? 64
2. 10 is 8% of what number? 125
3. 160% of what number is 152? 95
4. 30% of what number is 6? 20
5. 24 is 15% of what number? 160
6. 84 is 400% of what number? 21
7. 5% of what number is 6? 120
8. 22% of what number is 33? 150
9. 16 is 400% of what number? 4
10. 21 is 42% of what number? 50
11. 99 is 66% of what number? 150
12. 100% of what number is 300? 300
13. 4% of what number is 10? 250
14. 30% of what number is 150? 500
15. 50% of what number is 7.6? 15.2
16. 138 is 75% of what number? 184
17. 23 is 20% of what number? 115
18. 33% of what number is 33? 100
19. 125% of what number is 30? 24
20. 150 is 50% of what number? 300
21. 225 is 75% of what number? 300
22. 99% of what number is 198? 200
23. 3% of what number is 6? 200
24. 30% of what number is 5.4? 18
25. 8.5 is 20% of what number? 42.5
26. 15 is 60% of what number? 25
27. 80% of what number is 16? 20
28. 3 is 15% of what number? 20

Mixed Applications

29. At the school football game, 180 seats are empty. This is 40% of the seating capacity. What is the seating capacity?

    450 people

30. Mr. Hauser bought a suit on sale for 25% off the regular price. If he paid $120 for the suit, what was the regular price?

    $160

31. The area of a 20-ft-wide garden is 760 ft^2. How long is the garden?

    38 ft

32. A square and an equilateral triangle have the same perimeter. If a side of the square is 42 cm long, what is the length of a side of the triangle?

    56 cm

Name ______________________

LESSON 16.1

Similar Figures and Scale Factors

Vocabulary

1. Similar figures are figures that have the same ___shape___ but not necessarily the same ___size___.
2. When the pairs of corresponding sides of two figures have lengths that form equivalent ratios, you can say that the sides of the figures are ___proportional___.
3. What is the name of the common ratio of pairs of corresponding sides of similar figures? ___scale factor___

Determine whether the figures are similar. Write *yes* or *no*, and support your answer.

4.

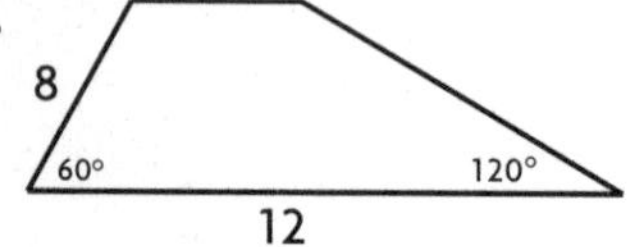

no; corresp. angles not congruent

5.

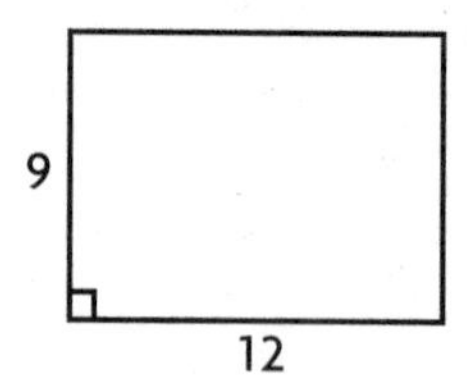

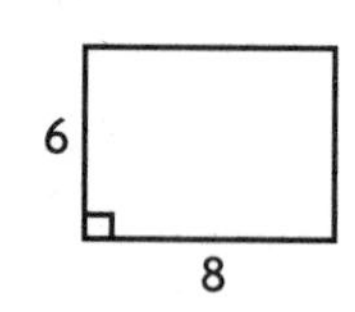

yes; angles congruent and $\frac{9}{6} = \frac{12}{8}$

The sides of triangle *ABC* are 3 cm, 4 cm, and 5 cm long. Draw a similar triangle, using the given scale factor. Give the length of each side. **Check students' drawings.**

6. scale factor: $\frac{1}{4}$

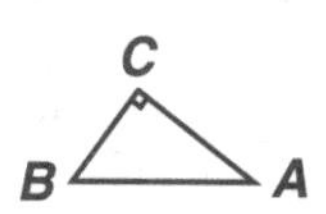

0.75 cm, 1 cm, 1.25 cm

7. scale factor: $\frac{6}{5}$

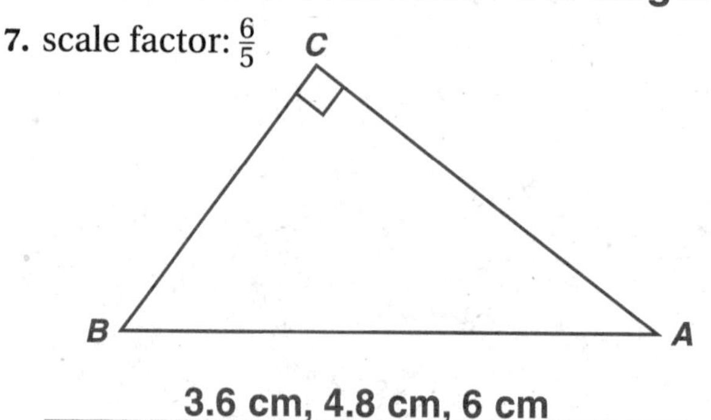

3.6 cm, 4.8 cm, 6 cm

Mixed Applications

8. Mary June has a 12-in. × 14-in. print that was made into a poster by using a scale factor of $\frac{9}{2}$. What is the area of the poster?

3,402 in.^{2}

9. Mrs. Smith's car averages 21.72 miles per gallon of gasoline. How many miles can she travel on 11.25 gallons of gasoline?

244.35 mi

Use with text pages 304–307.

Name ________________________

Proportions and Similar Figures

The figures in each pair are similar. Use a proportion to find the unknown length.

1.

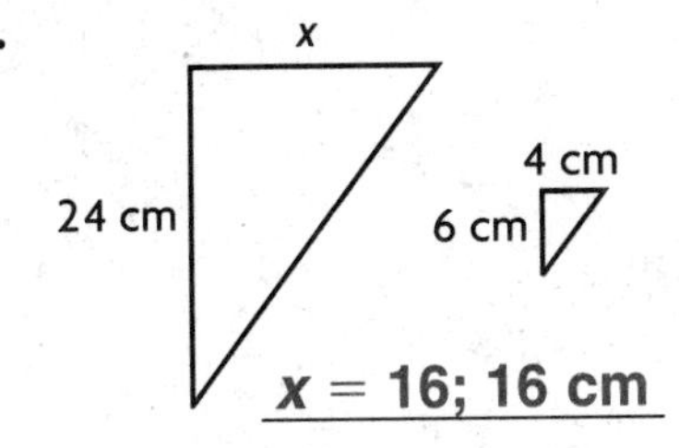

$x = 16$; 16 cm

2.

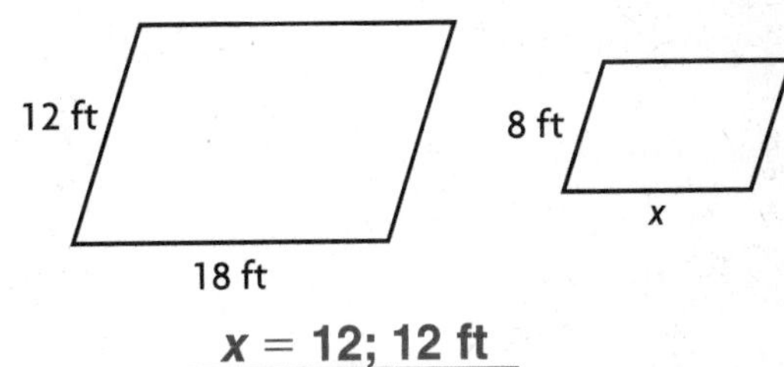

$x = 12$; 12 ft

3.

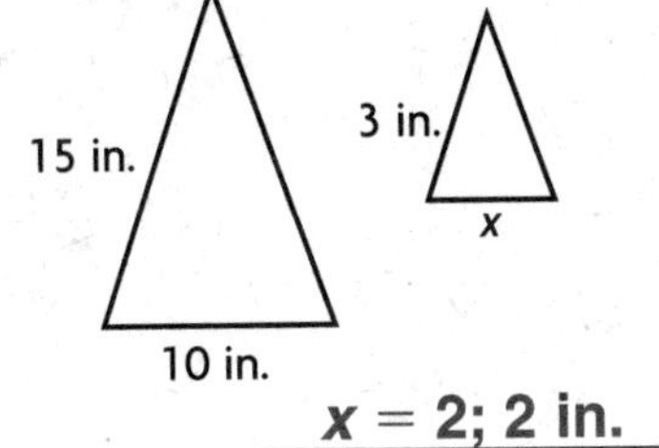

$x = 2$; 2 in.

4.

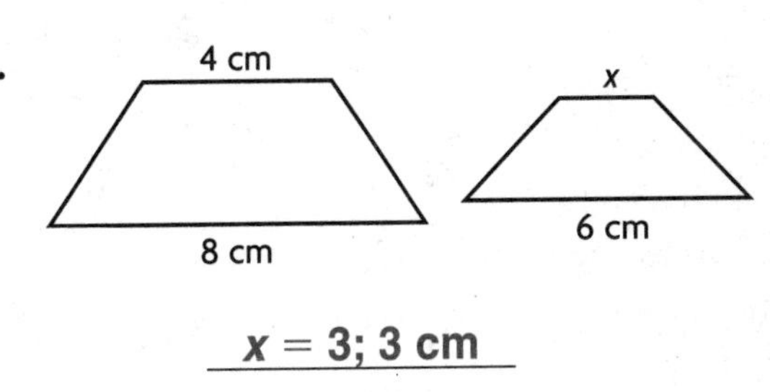

$x = 3$; 3 cm

Complete each table, using a scale factor of 3:1 for the enlarged model.

		Real Object	Enlarged Model
5.	Length	20 in.	60 in.
6.	Width	5 in.	15 in.
7.	Height	8 in.	24 in.

		Real Object	Enlarged Model
8.	Length	4.8 cm	14.4 cm
9.	Width	3.2 cm	9.6 cm
10.	Height	11.9 cm	35.7 cm

Mixed Applications

11. The number of students in this year's seventh-grade class is 110% of the number in last year's class. If there were 180 students in last year's class, how many are in this year's class?

198 students

12. Amy is making a clay sculpture of her hand in art class. First, she makes a scale drawing. She uses a 6:1 scale factor. If the drawing of Amy's hand is 9.5 cm wide, how wide will the sculpture be?

57 cm

13. On Sunday Scott spends $\frac{5}{12}$ of his time sleeping and $\frac{1}{8}$ of his time eating. How many more hours of the 24-hour period does Scott spend sleeping than eating?

7 hr more

14. An object's weight on Earth's surface is 6 times as much as on the moon's surface. If a man weighs 240 lb on Earth, what would he weigh on the moon?

40 lb

Name ______________________________

Areas of Similar Figures

Write a ratio to relate the area of the similar enlargement to the area of the original rectangle. The widths (w) and lengths (l) are given.

1. original rectangle: $w = 3$ in., $l = 4$ in.

   enlargement: $w = 9$ in., $l = 12$ in. **9 to 1**

2. original rectangle: $w = 2$ ft, $l = 5$ ft

   enlargement: $w = 3$ ft, $l = 7.5$ ft **2.25 to 1**

The scale factor for the sides of two similar polygons is given. Find the scale factor for their areas.

3. $\frac{4}{3}$ **$\frac{16}{9}$, or 16:9**
4. 7:3 **49:9, or $\frac{49}{9}$**
5. $\frac{6}{5}$ **$\frac{36}{25}$, or 36:25**
6. 9:2 **81:4, or $\frac{81}{4}$**

Find the areas of the similar polygons. Then find the scale factor for the areas. The formula for the area of a triangle is $A = \frac{1}{2}bh$. The formula for the area of a trapezoid is $A = \frac{1}{2}(b_1 + b_2)h$.

7.

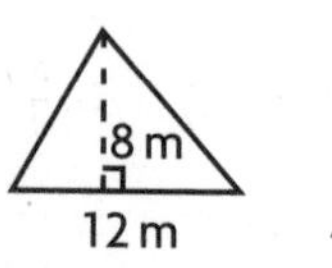

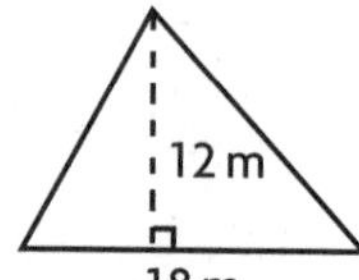

48 m²; 108 m²; $\frac{9}{4}$, or 2.25

8.

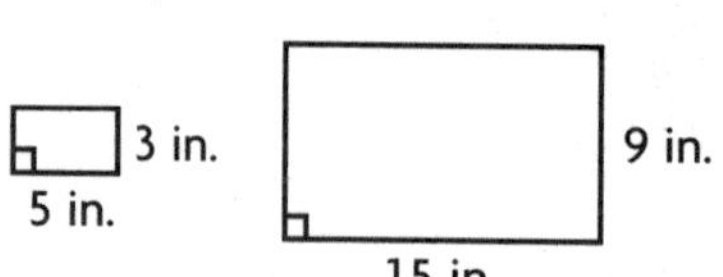

15 in.²; 135 in.²; $\frac{9}{1}$, or 9

9.

4 in.
2 in.
6 in.

8 in.
4 in.
12 in.

10 in.²; 40 in.²; $\frac{4}{1}$, or 4

Mixed Applications

10. Andrea wants to hang two similar posters on her wall. The dimensions of one poster are 10 in. × 12 in., and the dimensions of the other are 30 in. × 36 in. What scale factor relates the dimensions of the two posters? What scale factor relates the areas of the two posters?

    $\frac{1}{3}$; $\frac{1}{9}$

11. On October 1, Mark's savings account showed a balance of \$3,536.19. During the next month, he deposited \$432.12 and withdrew \$214.63. On November 1, his bank statement had a balance, including interest, of \$3,795.81. How much interest did he earn in October?

    \$42.13

12. Carlos has a sheet of paper that is 6 ft × 7 ft. On the paper he plans to make a design that consists of 36 8-in. × 10-in. rectangles. If he decides to use larger rectangles, using a scale factor of $\frac{3}{2}$ for the sides, can he fit the larger rectangles on the paper? Explain. **No. The larger rectangles would need 6,480 in.², and the paper is 6,048 in.²**

13. Mrs. Schneider buys 10 cases of juice for a school party. A case of juice contains 24 bottles. There are 6 oz of juice in each bottle. How many ounces of juice are in 10 cases?

    1,440 oz

Name ___

Volumes of Similar Figures

Tell the dimensions of a similar enlarged figure, made with a scale factor of 1.5 for the sides.

1.

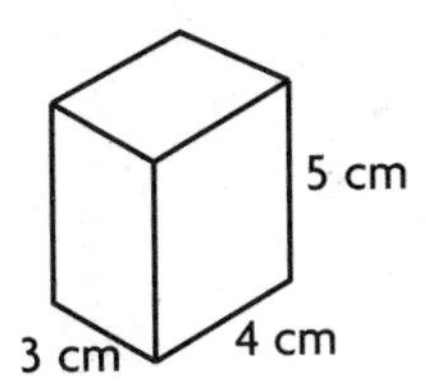

4.5 cm; 6 cm; 7.5 cm

2.

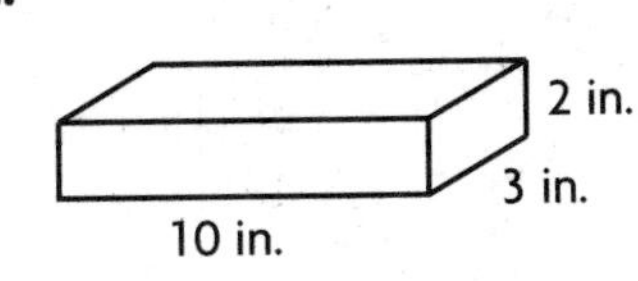

15 in.; 4.5 in.; 3 in.

3.

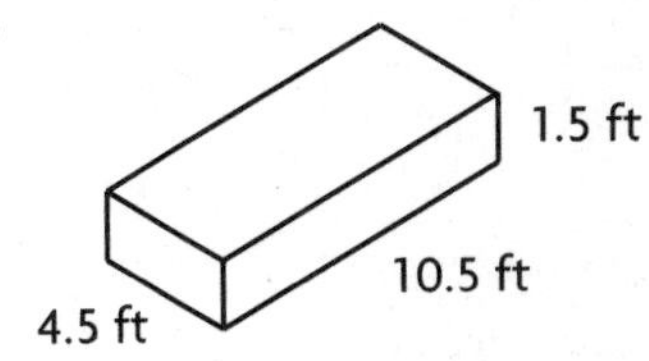

6.75 ft; 15.75 ft; 2.25 ft

Find the volumes of the similar prisms. Then find the scale factor for the volumes, relating the larger prism to the smaller prism.

4.

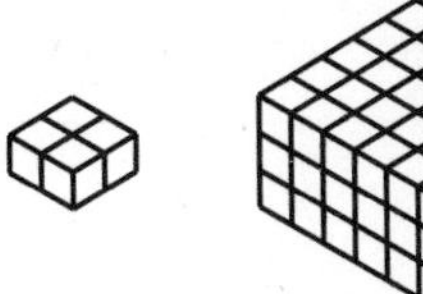

$V_{small} = 4$ units^3; $V_{large} = 108$ units^3;

scale factor: 27

5.

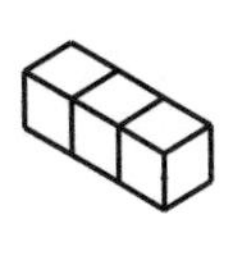

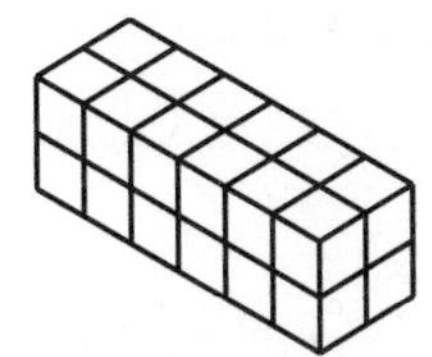

$V_{small} = 3$ units^3; $V_{large} = 24$ units^3;

scale factor: 8

Write a scale factor for the volumes, relating the larger cube to the smaller cube.

6.

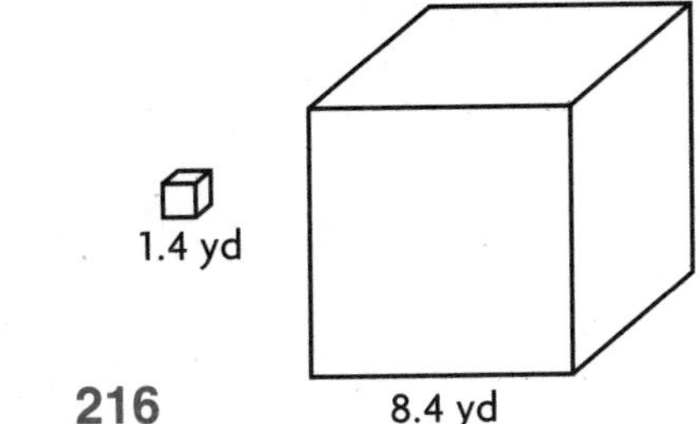

216

7.

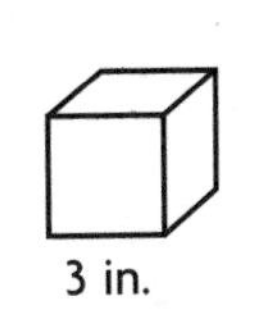

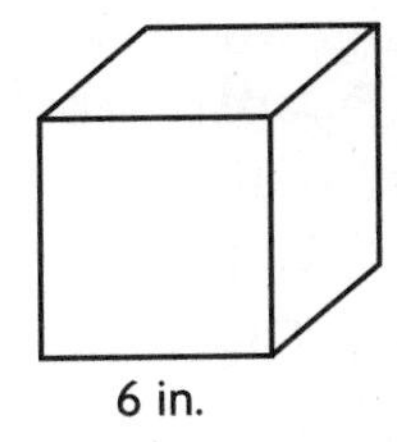

8

Mixed Applications

8. The dimensions of a small box of chocolates are 3 in. × 6 in. × 8 in. The small box sells for $4.32. The length, width, and height of a large box are each 2.5 times that of the small box. How much should the large box of chocolates cost?

$67.50

9. Peter Piper's pumpkin patch is 30 yd 1 ft × 72 yd 2 ft. What is the perimeter of Peter Piper's pumpkin patch?

206 yd

Name ___

LESSON 17.1

Drawing Similar Figures

Use the scale factor to draw a triangle that is similar to the given triangle. Write the lengths of the sides. **Check students' drawings.**

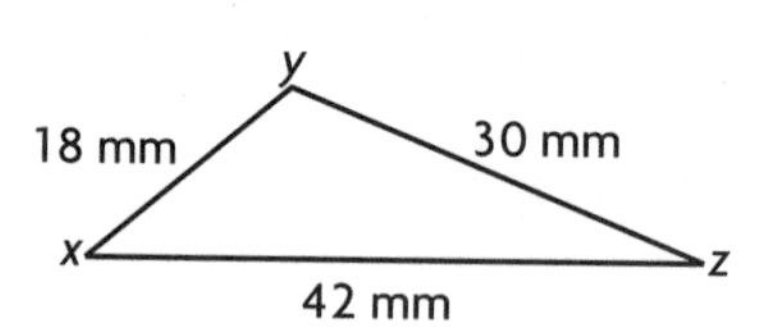

1. scale factor: $\frac{1}{3}$ **6 mm, 10 mm, 14 mm**
2. scale factor: 1.5 **27 mm, 45 mm, 63 mm**
3. scale factor: $\frac{1}{2}$ **9 mm, 15 mm, 21 mm**
4. scale factor: 2 **36 mm, 60 mm, 84 mm**

Use the scale factor to draw a rectangle that is similar to the given rectangle. Write the lengths of the sides. **Check students' drawings.**

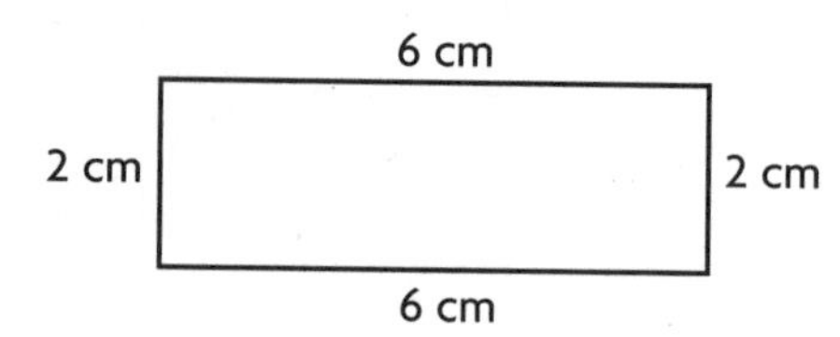

5. scale factor: $\frac{1}{2}$ **1 cm, 3 cm, 1 cm, 3 cm**
6. scale factor: 2.5 **5 cm, 15 cm, 5 cm, 15 cm**

Use the scale factor to draw a trapezoid that is similar to the given trapezoid. Write the lengths of the sides. **Check students' drawings.**

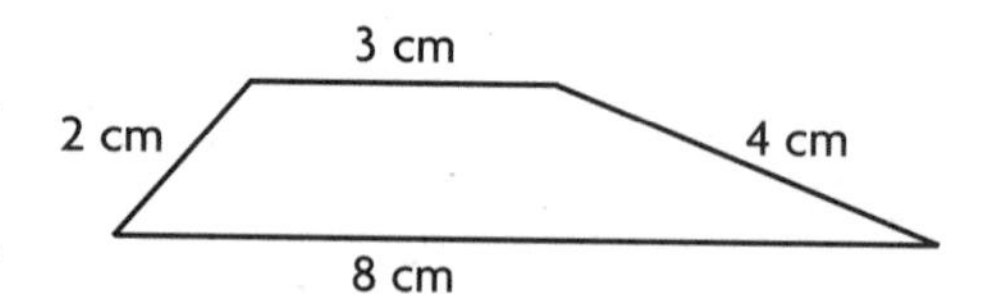

7. scale factor: $\frac{1}{4}$ **0.5 cm, 0.75 cm, 1 cm, 2 cm**
8. scale factor: 1.5 **3 cm, 4.5 cm, 6 cm, 12 cm**

Mixed Applications

9. Mark made a logo for a soup can label that was 6 in. × 18 in. He needs to enlarge the label by a scale factor of 1.5 for a larger can of soup. What are the dimensions of the label for the larger can of soup?

   9 in. × 27 in.

10. At Sam's Supermarket you can buy 7 apples for $1.89. At Gilda's Grocery you can buy 4 apples for $1.28. Where is the price per apple cheaper? Explain.

    Sam's Supermarket;

    1.89 ÷ 7 = $0.27,

    $1.28 ÷ 4 = $0.32; $0.27 < $0.32

11. Ira paid a total of $26.54 for a swimsuit. If he paid $1.59 in tax, what was the cost of the swimsuit?

    $24.95

12. At noon the temperature in Orlando is 88°F. It then drops 2°F per hour for 6 hours. What is the temperature at 6:00 P.M.?

    76°F

Use with text pages 322–323.

Name ______________________

Scale Drawings

Vocabulary

1. A scale drawing has the same **shape** as the object it represents.

2. The dimensions of a scale drawing are related by a **ratio**, or scale, to the dimensions of the actual object.

3. The ratio used to write the scale for scale drawings is $\frac{\text{drawing length}}{\text{actual length}}$.

Use the scale of 1 in.:25 in. to find the missing dimension. The first measure in the scale is for the scale drawing, and the second measure is for the actual object.

4. drawing: 2 in.
   actual: **50** in.

5. drawing: 3.5 in.
   actual: **87.5** in.

6. drawing: **0.5** in.
   actual: 12.5 in.

7. drawing: **1.4** in.
   actual: 35 in.

8. drawing: **2.1** in.
   actual: 52.5 in.

9. drawing: 4.2 in.
   actual: **105** in.

Use the scale of 25 in.:1 in. to find the missing dimension. The first measure in the scale is for the scale drawing, and the second measure is for the actual object.

10. drawing: **80** in.
    actual: 3.2 in.

11. drawing: 40 in.
    actual: **1.6** in.

12. drawing: 120 in.
    actual: **4.8** in.

Write an appropriate scale for a scale drawing of each item. Each scale drawing should fit on an $8\frac{1}{2}$-in. × 11-in. sheet of paper. **Possible answers are given.**

13. safety pin: 3 cm long
    10 in.:3 cm

14. telephone pole: 30 ft tall
    10 in.:30 ft or 1 in.:3 ft

15. car: 12 ft long
    10 in.:12 ft

Mixed Applications

16. Maxine is making a scale drawing of a gameboard for Field Day. She is using a scale of 1 in.:5 in. If the length of her drawing is 6 in., how many feet long is the actual gameboard?

    $2\frac{1}{2}$ ft

17. Oscar has $6 in his savings account and adds $1 each week. Alice has $10 in her savings account and adds $3 each week. After how many weeks will Alice have twice as much as Oscar?

    after 2 weeks

Name ____________________

Using Maps

Write and solve a proportion to find the actual distance using a map scale of 1 in.:15 mi.

1. map distance: 10 in. — 150 mi
2. map distance: $3\frac{1}{2}$ in. — $52\frac{1}{2}$ mi
3. map distance: 20 in. — 300 mi
4. map distance: $1\frac{1}{3}$ in. — 20 mi
5. map distance: $4\frac{1}{4}$ in. — $63\frac{3}{4}$ mi
6. map distance: $\frac{1}{4}$ in. — $3\frac{3}{4}$ mi
7. map distance: 4 in. — 60 mi
8. map distance: $5\frac{3}{4}$ in. — $86\frac{1}{4}$ mi
9. map distance: $2\frac{1}{2}$ in. — $37\frac{1}{2}$ mi
10. map distance: $\frac{1}{5}$ in. — 3 mi
11. map distance: 9 in. — 135 mi
12. map distance: $3\frac{3}{5}$ in. — 54 mi
13. map distance: $3\frac{2}{3}$ in. — 55 mi
14. map distance: $\frac{4}{5}$ in. — 12 mi
15. map distance: 5 in. — 75 mi
16. map distance: $\frac{1}{6}$ in. — $2\frac{1}{2}$ mi
17. map distance: $3\frac{1}{3}$ in. — 50 mi
18. map distance: $5\frac{2}{3}$ in. — 85 mi
19. map distance: $2\frac{1}{4}$ in. — $33\frac{3}{4}$ mi
20. map distance: $\frac{1}{3}$ in. — 5 mi
21. map distance: 8 in. — 120 mi

Mixed Applications

22. The Smythe family plans to drive from Rochester, NY, to Niagara Falls, NY, on their vacation. On the map, the distance is about $2\frac{1}{4}$ in. If the map scale is $\frac{1}{4}$ in.:16 mi, about how far will they be driving?

    144 mi

23. Sam and Spencer's trail map has a scale of 1 in.:2 mi. How far will they hike if the trail they want to use is 4.5 in. long on the map?

    9 mi

24. Marge wants to buy a CD player that costs $159.99. If the sales tax rate is $7\frac{1}{2}$%, how much money does she need?

    $171.99

25. A map has a scale of 2 cm:1 mi. What is the map distance for an actual distance of 4 mi? 10 mi? $2\frac{1}{2}$ mi?

    8 cm; 20 cm; 5 cm

26. It takes an 8-lb turkey $3\frac{1}{4}$ hr to cook. At this rate, how long will it take a 20-lb turkey to cook?

    $8\frac{1}{8}$ hr

27. If you paint four walls, each 8 ft tall and 12 ft wide, you will cover how many square feet?

    384 ft^2

Use with text pages 330–332.

Name ____________________

Indirect Measurement

1. Identify the similar triangles. a and d

a.

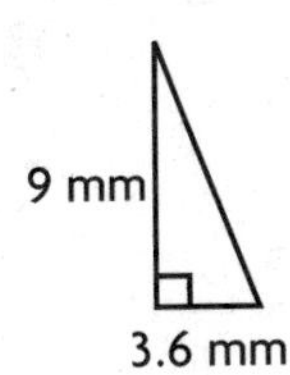

b.

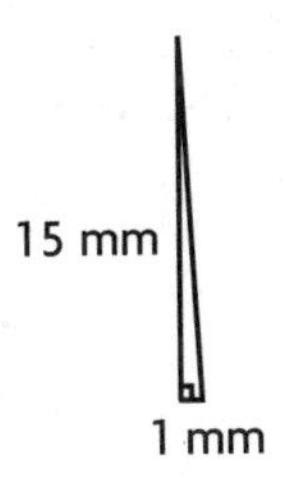

c.

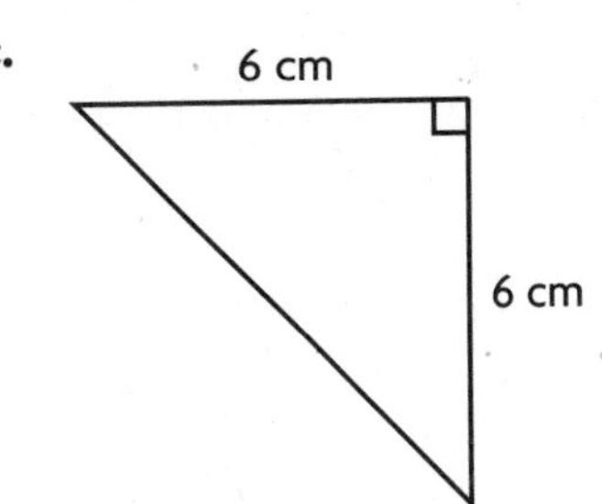

d.

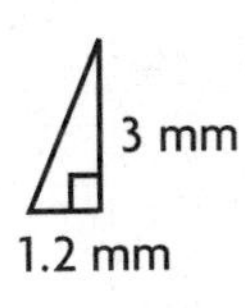

The triangles in each pair are similar. Find x.

2. $x = 9$ cm

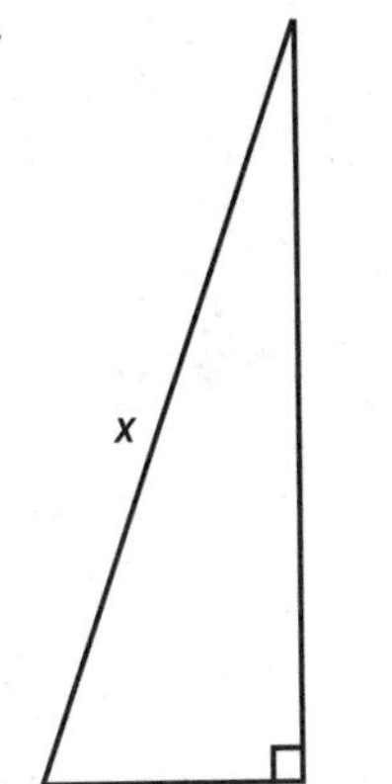

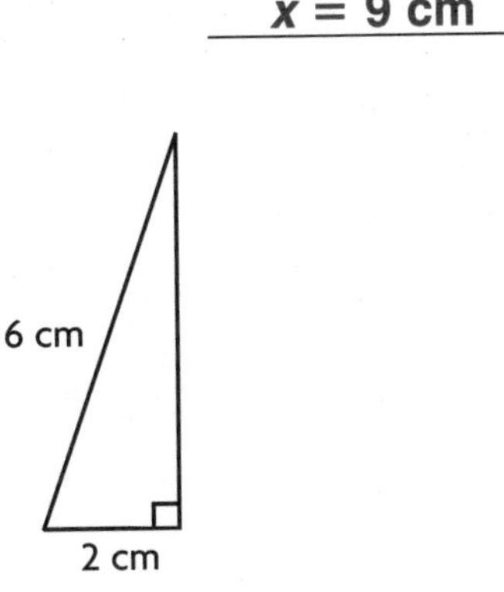

3. $x = 120$ mm

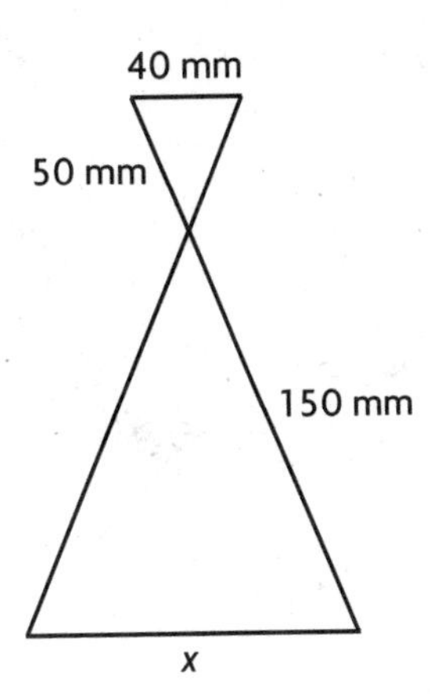

4.

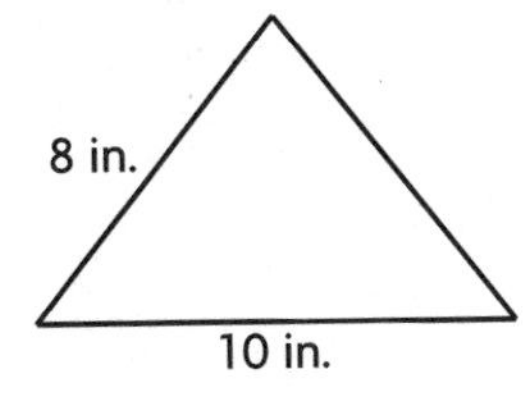

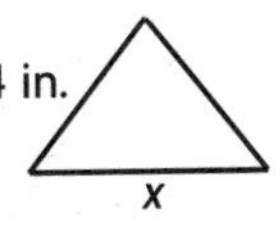

$x = 5$ in.

5.

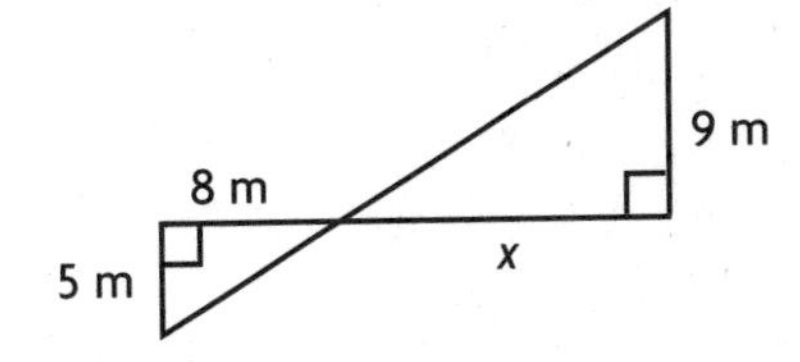

$x = 14.4$ m

6. A 120-m cable TV tower casts a 96-m shadow. Find the height of a nearby telephone pole that casts a 16-m shadow. Use a diagram of two similar triangles to help you solve.

20 m

7. A boy who is 180 cm tall casts a 160-cm shadow. A nearby flagpole casts a 10-m shadow. How tall is the flagpole? Use a diagram of two similar triangles to help you solve.

11.25 m

8. Max spent \$27.60 for dinner at a restaurant, including a \$3.60 tip. How much did the dinner cost without the tip?

\$24

9. The sum of two numbers is 60. Their difference is 4. What are the two numbers?

28 and 32

Name ______________________

Golden Rectangles

Vocabulary

In Exercises 1–2, select the correct answer by circling a, b, or c.

1. A Golden Rectangle is a rectangle with a length-to-width ratio of approximately
   a. 1.57 to 1. **(b. 1.61 to 1.)** **c.** 6.16 to 1.
2. A ratio is a Golden Ratio if it has a decimal value of approximately
   a. 1.57. **(b. 1.61.)** **c.** 6.16.

Is the rectangle a Golden Rectangle? Explain.

3. Yes; ratio of length to width is near 1.61.

   l = 12 cm, w = 7.5 cm

4. No; ratio of length to width is 1.75, rather than about 1.61.

   l = 28 cm, w = 16 cm

5. Yes; ratio of length to width is near 1.61.

   l = 6.3 cm, w = 3.9 cm

   New York Drivers License
   John Doe
   111 Anystreet
   New York, New York 12345
   License Number:
   098766543321
   Birth Date
   1/1/00
   Sex HT WT Hair Eyes
   Male 5-08 135 Green Red

6. Yes; ratio of length to width is near 1.61.

   l = 4, w = 2.5

Draw rectangles with the following dimensions. Are the rectangles Golden Rectangles? Support your answers.

7.

Length	Width
2.6 in.	1.6 in.

yes: $\frac{2.6}{1.6} = 1.625$

8.

Length	Width
8 cm	4.7 cm

no: $\frac{8}{4.7} = 1.7$

Mixed Applications

9. A painting shows a building with length 14 cm and width 8.5 cm. Is the building a Golden Rectangle? Explain.

   Yes; ratio $\frac{14}{8.5}$ is near 1.61.

10. Joe drove to his grandparents' house 496 miles away. He drove for 8 hr without stopping. At what average speed did he travel?

    62 mph

Use with text pages 336–337.

Name ______________________

LESSON 18.1

Triangular Arrays

Vocabulary

1. Write *true* or *false.* A triangular number can be shown geometrically with a square array. false

---

Find the triangular number.

2. seventh 28
3. eighth 36
4. twelfth 78
5. twentieth 210

For Exercises 6–7, use the array at the right.

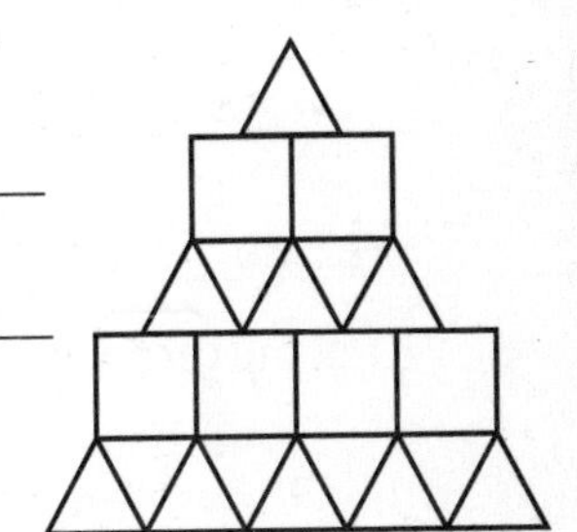

6. What rule was used to determine which figures are triangles and which are squares? triangles for odd-numbered rows, squares for even

7. What will be the next row in the array? 6 squares

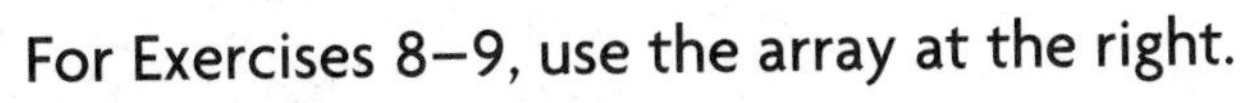

For Exercises 8–9, use the array at the right.

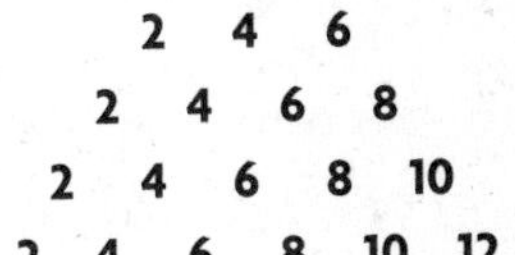

8. What will be the next row in the array? What pattern did you use to find the next row? 2 4 6 8 10 12 14; including the next even number at the end

9. How many numbers will be in Row 8? in Row 12? in Row *n*? 8; 12; *n*

Mixed Applications

10. The cheerleaders at Salem Middle School want to make a human pyramid 5 levels high. How many people will they need for this triangular array?

    15 people

11. A video store rents 7 children's videos for every 4 teenagers' videos. If it rents 24 videos for teenagers, how many children's videos does it rent?

    42 children's videos

12. Bob and John have 28 coins. They want to make a triangular array with the coins. How many rows will be in their array?

    7 rows

13. Hideko's normal pulse rate is 65 beats per minute. When she rides her bike, her pulse rate increases by 45 beats per minute. How many times does her heart beat during a 25-minute bike ride?

    2,750 times

Name ______________________

LESSON
18.2

Pascal's Triangle

Vocabulary

1. What is the name of a triangular array in which each row starts and ends with a 1 and each of the other numbers is the sum of the two closest numbers above it? Pascal's triangle

---

For Exercises 2–5, use the Pascal's triangle at the right.

Row 0 → 1
1 1 ← Row 1
1 2 1
1 3 3 1
1 4 6 4 1
1 5 10 10 5 1
1 6 15 20 15 6 1
1 7 21 35 35 21 7 1
1 8 28 56 70 56 28 8 1
1 9 36 84 126 126 84 36 9 1

2. What numbers would be in Row 10? How did you find these numbers? 1, 10, 45, 120, 210, 252, 210, 120, 45, 10, 1; by starting and ending the row with 1 and adding the two closest numbers above to find each other number.

3. What pattern do you find for the number of entries in each row as you move down the array? Each row has one more entry than the previous row.

4. What is the second number in Row 5? in Row 6? in Row 7? What will be the second number in Row 20? in Row 67? in Row 89? 5; 6; 7; 20; 67; 89

5. What do you notice about the numbers in the third diagonals (1, 3, 6, . . .)?

They increase by the counting numbers 2, 3, 4, 5, 6, . . .

Mixed Applications

6. Dolly's Frozen Yogurt Parlor has 5 sundae toppings. How many different sundaes can be made with 0, 1, 2, 3, 4, or 5 toppings?

32 different sundaes

7. Your city park had registration for fall hockey. The number of players registered was divided by 15 to form 12 teams. How many players signed up for hockey?

180 players

8. Chef Charlie is making tacos. There are 6 extra toppings. How many varieties of tacos can Chef Charlie make with 0, 1, 2, 3, 4, 5, or 6 extra toppings?

64 varieties of tacos

9. The price of a circus ticket is $6 at the door or $4 in advance. What is your percent savings if you buy tickets in advance?

$33\frac{1}{3}\%$

Name ______________________

Repeated Doubling and Halving

Vocabulary

1. If you keep halving a positive number, do the numbers converge or diverge? converge
2. If you keep doubling a positive number, do the numbers converge or diverge? diverge

---

Determine how many times 6 must be doubled to reach or exceed the given number.

3. 192 — 5
4. 3,072 — 9
5. 24,576 — 12
6. 393,216 — 16

Determine how many times 250,000 must be halved to reach or be less than the given number.

7. 31,250 — 3
8. 980 — 8
9. 245 — 10
10. 16 — 14

Tell whether the numbers converge or diverge.

11. $7, 3\frac{1}{2}, 1\frac{3}{4}, \frac{7}{8}, \ldots$ converge
12. 15, 30, 60, 120, . . . diverge

For Exercises 13–14, use the graph.

13. Do the data displayed in the graph converge or diverge? converge
14. Suppose the number of students taking school music lessons continues to halve. In what year will 100 students be taking school music lessons? 1998

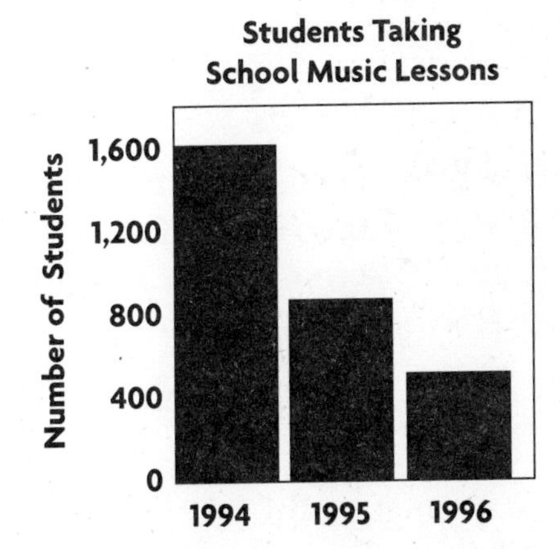

Mixed Applications

15. A pen contains thirty rabbits. If the rabbit population doubles every 6 months, how many rabbits will there be after 3 years?

1,920

16. José plants vegetables in $\frac{1}{4}$ of his garden, flowers in $\frac{1}{3}$, and berries in the rest. How much of his garden is planted with berries?

$\frac{5}{12}$ of his garden

Name ______

LESSON 18.4

Exponents and Powers

Use the diagram to find the first five powers.

1. Start with 1. → Multiply by 11. (repeat)

11; 121; 1,331; 14,641; 161,051

2. Start with 1. → Multiply by $\frac{2}{3}$. (repeat)

$\frac{2}{3}, \frac{4}{9}, \frac{8}{27}, \frac{16}{81}, \frac{32}{243}$

For Exercises 3–6, find the first four powers of the given number.

3. 3 **3, 9, 27, 81**

4. $\frac{1}{5}$ $\frac{1}{5}, \frac{1}{25}, \frac{1}{125}, \frac{1}{625}$

5. 20 **20; 400; 8,000; 160,000**

6. 8 **8; 64; 512; 4,096**

Find the values. Then show the sequence as an orbit of a point. **Check students' orbits.**

7. $6^0, 6^1, 6^2, 6^3$ **1, 6, 36, 216**

8. $\left(\frac{1}{10}\right)^0, \left(\frac{1}{10}\right)^1, \left(\frac{1}{10}\right)^2, \left(\frac{1}{10}\right)^3$ $1, \frac{1}{10}, \frac{1}{100}, \frac{1}{1{,}000}$

Find the exponent n that will make the equation true.

9. $5^n = 125$ **3**

10. $10^n = 1{,}000{,}000$ **6**

11. $12^n = 144$ **2**

Mixed Applications

12. Daniel's gerbil population quadruples every 3 months. He has 10 gerbils now. How many gerbils will he have after 1 year?

2,560 gerbils

13. Rosa made a $15\frac{1}{2}$-in. × $6\frac{7}{8}$-in. poster for art class. How much longer is the poster than it is wide?

$8\frac{5}{8}$ **in.**

14. The original size of your brother's graduation picture is 8 in. × 10 in. You copy it 3 times. Each time you copy it, its dimensions are reduced by $\frac{1}{2}$. What is the picture's final size?

1 in. × $1\frac{1}{4}$ **in.**

15. Roger's math class covers 2 chapters in 21 school days. At this rate, how many school days will it take to cover 15 chapters?

$157\frac{1}{2}$ **school days**

Name ______________________________

Exploring Patterns in Decimals

Vocabulary

Complete.

1. When you divide to change a fraction to a decimal and the remainder is not zero, the quotient is a(n) **repeating decimal**.

For Exercises 2–5, choose the equivalent decimal from the box.

0.4	$0.\overline{6}$	$2.\overline{1}$	$1.8\overline{3}$

2. $\frac{2}{3}$ **$0.\overline{6}$**
3. $\frac{19}{9}$ **$2.\overline{1}$**
4. $\frac{11}{6}$ **$1.8\overline{3}$**
5. $\frac{2}{5}$ **0.4**

Write *R* if the fraction can be renamed as a repeating decimal. Write *T* for a terminating decimal. Write the decimal equivalent, using a bar for a repeating decimal.

6. $\frac{7}{8}$ **T; 0.875**
7. $\frac{7}{12}$ **R; $0.58\overline{3}$**
8. $\frac{7}{6}$ **R; $1.1\overline{6}$**
9. $\frac{13}{11}$ **R; $1.\overline{18}$**

For Exercises 10–13, rewrite the fraction as a decimal. Do not divide. Use the values $\frac{1}{9} = 0.\overline{1}$, $\frac{1}{8} = 0.125$, and $\frac{1}{3} = 0.\overline{3}$.

10. $\frac{5}{8}$ **0.625**
11. $\frac{7}{9}$ **$0.\overline{7}$**
12. $\frac{4}{3}$ **$1.\overline{3}$**
13. $\frac{13}{9}$ **$1.\overline{4}$**

Mixed Applications

14. Scott buys a large pizza, which is cut into 16 slices. He eats 5 slices. Find the decimal value of the remaining portion of the pizza.

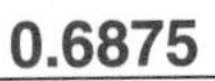

0.6875

15. Dean's Dress Shop advertised a savings of \$10 for every \$60 spent. How much would Anna have to spend to save \$90?

\$540

16. Allie walks $2\frac{7}{12}$ mi to school each day. Find the decimal equivalent of $2\frac{7}{12}$.

$2.58\overline{3}$

17. A class trip to the zoo costs \$92.75. The admission fee is \$2.75 for students and \$4.25 for adults. Five adults go on the trip. How many students go?

26 students

Name ____________________

LESSON 19.2

Patterns in Rational Numbers

Vocabulary

1. Choose the best answer. The Density Property states that between any two rational numbers there is __b__.

   a. an integer b. a rational number c. a whole number d. a negative number

Name a rational number for the given point on the number line.

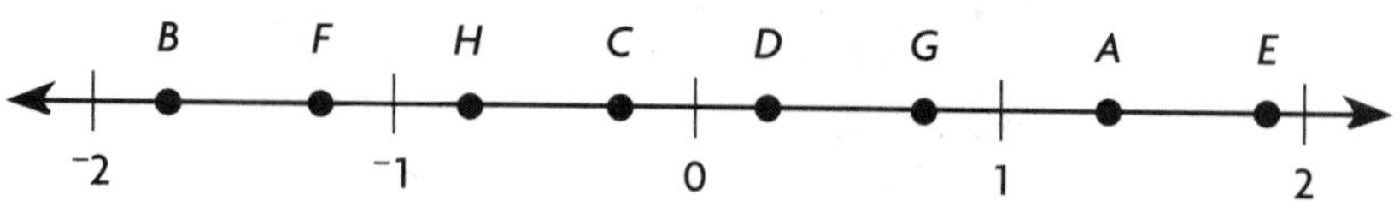

2. point A — $1\frac{3}{8}$
3. point B — $-1\frac{3}{4}$
4. point C — $\frac{-1}{4}$
5. point D — $\frac{1}{4}$

For the given rational number, name the point on the number line above.

6. $\frac{-3}{4}$ — H
7. $^{-}1\frac{1}{4}$ — F
8. $1\frac{7}{8}$ — E
9. 0.75 — G

Name a rational number between the two numbers. **Possible answers are given.**

10. 4.03 and 4.04 — 4.035
11. 1.62 and 1.64 — 1.63
12. $\frac{1}{5}$ and $\frac{3}{5}$ — $\frac{2}{5}$
13. $^{-}1$ and 0 — $\frac{-1}{2}$
14. 1.5 and $1\frac{3}{8}$ — 1.4
15. $\frac{5}{10}$ and $\frac{7}{10}$ — $\frac{6}{10}$
16. 50.1 and 50.15 — 50.125
17. $^{-}4\frac{1}{2}$ and $^{-}4$ — $-4\frac{1}{4}$

Mixed Applications

18. Mark was not feeling well, so he took his temperature. The first reading was 96.4°F and the second reading was 101.2°F. His normal temperature is halfway between the two readings. What is Mark's normal temperature?

    98.8°F

19. The scale on a road map of Pennsylvania is $\frac{1}{4}$ in.:10 mi. The distance from Pittsburgh to Philadelphia on the map is about $7\frac{1}{4}$ in. Find the actual distance in miles between the two cities.

    about 290 mi

20. Marge plays center for her school basketball team. In her last game she made 75% of 16 free throws. How many free throws did she make?

    12 free throws

21. The students in science class are measuring the width of a classroom desk. The measurements are between $15\frac{1}{2}$ in. and $15\frac{7}{8}$ in. What might be the actual measurement of the desk?

    Possible answers: $15\frac{3}{4}$ in. or $15\frac{5}{8}$ in.

Use with text pages 365–367.

Name ______________________

Patterns in Sequences

Vocabulary

Complete.

1. When the pattern in a sequence is made by multiplying by the same number, the sequence is called a(n) geometric sequence.
2. The number used to multiply each term in a geometric sequence is called the common ratio.

Look for the pattern in each sequence. Write *geometric, arithmetic,* or *neither.*

3. 5, 12, 19, 26, . . . arithmetic
4. 3, 6, 10, 15, . . . neither
5. 34, 31, 28, 25, . . . arithmetic
6. 2, 6, 18, 54, . . . geometric

Write the next two terms in the sequence.

7. 3, 6, 12, 24, . . . 48; 96
8. 5, 8, 11, 14, . . . 17; 20
9. 12, 36, 108, . . . 324; 972
10. 72, $^-36$, 18, . . . $^-9$; 4.5
11. 7, 8, 10, 13, . . . 17; 22
12. $^-5, ^-1, \frac{^-1}{5}, \frac{^-1}{25}, \ldots$ $\frac{^-1}{125}; \frac{^-1}{625}$
13. 11, 17, 23, 29, . . . 35; 41
14. 0.875, 1.75, 3.5, 7, . . . 14; 28

Mixed Applications

15. At the grocery store, you see a display of soup cans in the shape of a pyramid. There are 6 cans in the top row, 8 cans in the next row down, 10 in the next row, and so on. How many cans of soup are in the sixth row?

    16 cans

16. You work for Fedler's Department Store, which is having a 30%-off sale on jackets. Employees receive an additional 20% off the sale price. How much would you pay for a jacket that originally cost $60?

    $33.60

17. You invest $200 in stock that increases 1.25 times in value each year. What will the value of your stock be at the end of the fifth year?

    $610.35

18. A rocket rises 30 ft in the first second after take-off, 60 ft in the second second, and 90 ft in the third second. If this pattern continues, how many feet will it rise in the eighth second?

    240 ft

Use with text pages 369–371.

Name ____________________

Patterns in Exponents

Write each expression, using a positive exponent.

1. 5^{-3} $\frac{1}{5^3}$
2. 2^{-1} $\frac{1}{2^1}$
3. 11^{-4} $\frac{1}{11^4}$
4. 10^{-2} $\frac{1}{10^2}$
5. 7^{-5} $\frac{1}{7^5}$
6. 4^{-2} $\frac{1}{4^2}$
7. 3^{-6} $\frac{1}{3^6}$
8. 10^{-7} $\frac{1}{10^7}$

Write each expression, using a negative exponent.

9. $\frac{1}{5^5}$ 5^{-5}
10. $\frac{1}{6^2}$ 6^{-2}
11. $\frac{1}{7^1}$ 7^{-1}
12. $\frac{1}{4^6}$ 4^{-6} or 2^{-12}
13. $\frac{1}{10^3}$ 10^{-3}
14. $\frac{1}{1,000,000}$ 10^{-6}
15. $\frac{1}{3^4}$ 3^{-4}
16. $\frac{1}{4 \times 4 \times 4}$ 4^{-3} or 2^{-6}
17. $\frac{1}{125}$ 5^{-3}

Write *positive exponent* or *negative exponent* to tell how you would express the given number.

18. Solar flares travel through space at 3,000,000 km/hr. positive exponent

19. A computer can do addition in 0.00000014 sec. negative exponent

Mixed Applications

20. In a 50-yd freestyle race, Art came in first, with a time of 36.24 sec. Bill's second-place time was $\frac{1}{100}$ sec greater than Art's time. Write the difference between their times, using a negative exponent. What was Bill's time?

10^{-2}; 36.25 sec

21. At the Metro Theater, general admission tickets to a concert cost $8 and student tickets cost $3. Mario sold 20 tickets and collected $100. How many student tickets were sold?

12 student tickets

22. A millimeter is $\frac{1}{1,000}$ of a meter. A kilometer is 1,000 m. How many millimeters are in a kilometer? Write the number as an expression with a positive exponent.

10^6 mm

23. Mr. Kurt just started working at Freshest Grocery Store. His starting wages are $6 per hour. Every 6 months, he will earn a raise of $0.50 per hour. What will his wages be after 3 years?

$9 per hr

Use with text pages 372–373.

Name ______________________

Choosing a Sample

Vocabulary

Write the correct letter from Column 2.

1. population ___d___
2. random sample ___a___
3. sample ___e___
4. stratified sample ___b___
5. systematic sample ___c___

a. sampling method in which every individual or object in a given population has an equal chance of being selected
b. sampling method in which a population is divided into subgroups, called strata, that contain similar individuals or objects
c. sampling method in which you randomly select an individual or object, and then follow a pattern to select others
d. whole group of people or objects
e. part of a group

---

6. Rico is conducting a survey to find out how many students own a neck-lace. How can he get a systematic sample?

Possible answer: by surveying every fifth person

7. The following table shows the results when people were randomly selected to name their favorite holiday. Show how the results might look for a stratified sample. **Tables will vary.**

Random Sample	
Valentine's Day	Fourth of July
Halloween	Thanksgiving
Veteran's Day	New Year's Day

Stratified Sample	

Indicate the type of sample.

8. Ten teenagers and ten senior citizens are asked if they prefer fiction or nonfiction books.

stratified sample

9. Students whose I.D. numbers contain a zero are asked to identify their favorite teacher.

systematic sample

Mixed Applications

10. Could a sample of 100 seventh-grade girls and 100 seventh-grade boys fairly represent your school population? Explain.

Answers will vary.

11. The map distance between Centerville and Riverton is $8\frac{1}{2}$ in. If the scale is 1 in. equals 60 mi, what is the actual distance between Centerville and Riverton?

510 mi

Name ______________________

Bias in Samples

Vocabulary

Complete.

1. A sample is __biased__ if any individual in the population is not proportionately represented by the sample.

Answer the following questions.

2. Is a survey of teenagers biased if it includes a senior citizen?

   Yes; a senior citizen is not part of the given population.

3. Is a survey of favorite baseball gloves biased if it encourages you to respond a certain way? Yes; it directs a possible response.

An ice cream manufacturer surveys 500 people about the flavor of ice cream they like. Tell whether the given sample is biased. If it is, tell why.

4. A random survey of 200 people

   not biased

5. A random survey of 200 school-age children

   biased because no adults are surveyed

Determine whether the question is biased.

6. Do you agree that bananas are the best fruit?

   biased

7. What is your favorite fruit?

   not biased

Mixed Applications

8. Give an example of a biased question.

   Possible response: "Do you think R.L. Stine is the best author?"

9. What is the next term in the following sequence? 0.13, 0.18, 0.28, 0.43, 0.63, . . .

   0.88

10. Change your question for Exercise 8 to make it unbiased.

    Possible response: "Who is your favorite author?"

11. A 45-g serving of Molly's favorite cereal contains 18 g of sugar. What percent of the cereal is sugar?

    40%

Use with text pages 385–386.

Name ______________________

LESSON 20.3

Writing Survey Questions

For Exercises 1–8, name the format of the survey question.

1. Using a scale of 1 to 5, how would you rate frozen raspberry yogurt?

   numerical

2. What is your favorite flavor of frozen yogurt?

   short answer

3. What is your favorite gym activity?

   short answer

4. Your favorite frozen yogurt topping is fill-in-the-blank.

5. Using a scale of 1 to 5, how would you rate television as entertainment?

   numerical

6. What is your favorite activity?

   short answer

7. Which is your favorite activity?

   a. reading **b.** watching TV
   c. playing sports **d.** listening to music

   multiple choice

8. From 1 to 10, how would you rate reading as a leisure activity?

   numerical

Tell whether the question would be a good survey question. Write *yes* or *no.* If you write *no,* explain.

9. Is your favorite sport football, soccer, or tennis?

   No; it is biased.

10. What type of music do you prefer?

    yes

11. Should the long soccer team practices last more than two hours?

    No; it is biased.

12. What should the soccer coach do to improve the quality of the team?

    No; it requires a long answer.

Mixed Applications

13. Write a survey question in numerical format about a television show.

    Answers will vary.

14. A neon sign blinks once every 5 sec. How often does the sign blink in 1 hr?

    720 times

15. It takes James $4\frac{1}{2}$ days to drive 1,800 mi. Driving at the same rate, how long would it take him to drive 3,000 mi?

    $7\frac{1}{2}$ days

16. Write a multiple-choice survey question about seventh-grade students' favorite movies.

    Answers will vary.

Name ______________________________

LESSON 20.4

Organizing and Displaying Results

1. Are the tens digits or the ones digits the stems in a stem-and-leaf plot of two-digit whole numbers?

tens digits

2. List the math test scores shown in this line plot.

82, 84, 88, 88, 91, 94, 97, 97, 100, 100

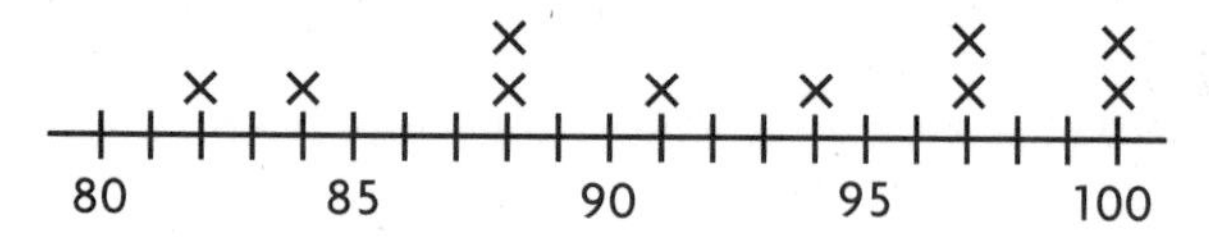

3. Name the intervals in the stem-and-leaf plot.

40–49, 50–59,

60–69, 70–79

Stem	Leaves
4	0 6
5	0 7 8
6	2 4
7	0 8 9 9

6|2 represents 62

Seventh graders were asked about their favorite team sports. The results are in the table.

SEVENTH GRADERS' FAVORITE SPORTS			
Team Sport	Tally	Frequency	Cumulative Frequency
Baseball	𝍸 ll	7	7
Basketball	𝍸 l	6	13
Football	𝍸 𝍸 ll	12	25
Soccer	𝍸 lll	8	33

4. Which team sport was the most popular? How many students liked it?

football; 12 students

Mixed Applications

5. Make a stem-and-leaf plot for the following students' heights: 137 cm, 144 cm, 132 cm, 148 cm, 151 cm, 149 cm, 155 cm, 134 cm, 144 cm, 154 cm, 150 cm.

Check students' work.

6. Make a cumulative frequency table to show that in Mr. Number's math class 8 students prefer algebra, 5 prefer geometry, 3 prefer statistics, and 7 prefer probability. What is the cumulative frequency for all the types of math? What is the cumulative frequency for statistics and probability?

Check students' tables; 23, 10.

7. Find the mean, median, and range of these temperature readings: 99°F, 83°F, 88°F, 77°F, 100°F, 91°F, 85°F.

89°F; 88°F; 23°F

8. The difference between the digits in a two-digit number is 5. If you add the digits and divide by 3, you get 3. What is the number?

27 or 72

Use with text pages 390–393.

Name ______________________

How Do Your Data Shape Up?

For Exercises 1–4, use the stem-and-leaf plot shown here.

Math Test Scores (in %)	
Stem	Leaves
5	3 5 8
6	2 5 6 9
7	0 5 8 8 8 9
8	0 2 4 5 5
9	3 5 8

1. What is the lowest test score? the highest test score? the range of scores? **53%; 98%; 45%**

2. In what interval, shown as a stem, do most of the scores occur? **70–79%**

3. Were there more test scores in the sixties or in the eighties? **eighties**

4. Were more than half the scores less than 78? Explain. **No; only 9 of 21 scores were less than 78.**

For Exercises 5–6, use the box-and-whisker graph below, which shows the distribution of money spent at a grocery store.

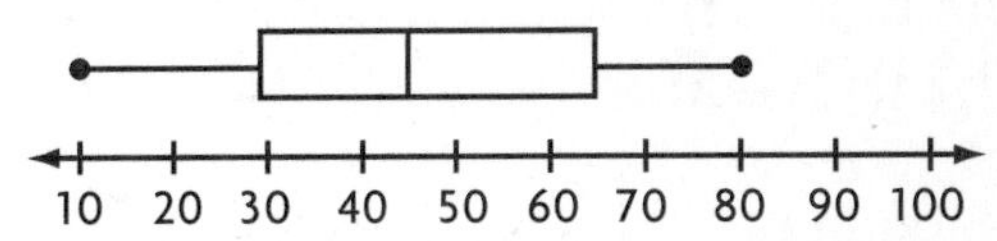

5. What is the median amount spent? What are the greatest and least amounts spent? What is the range of money spent?

$45; $80 and $10; $70

6. Do most of the shoppers spend more than $30? Explain.

Yes; only 25% of the shoppers spend less than $30.

Mixed Applications

7. Construct a histogram for the data at the right.

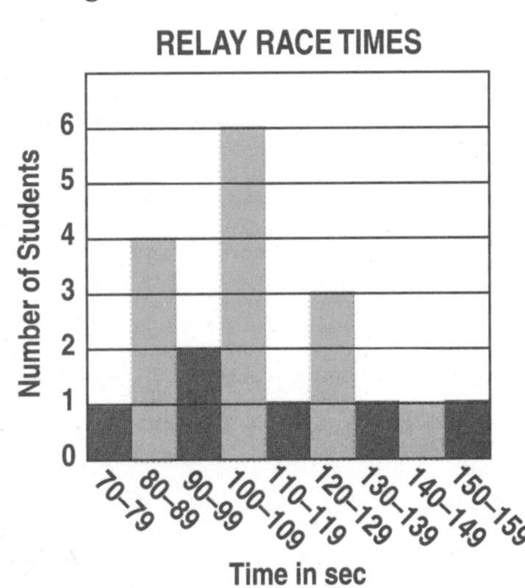

Relay Race Times (in seconds)			
70	105	121	101
130	100	114	102
150	100	142	90
80	100	82	94
80	126	85	127

8. Mark's rate of reading is 22 pages in 30 min. At this rate, how long will it take him to read 121 pages?

165 min

Name ______________________

LESSON 21.2

Central Tendencies

Vocabulary

Complete.

1. A single number that represents "the middle" of a set of data is called a

   measure of **central tendency**.

Find the mean, median, and mode of the set of numbers.

2. 43, 43, 55, 43, 54, 46, 56, 52

   mean: 49; median: 49; mode: 43

3. 61, 89, 93, 102, 47, 93, 61

   mean: 78; median: 89; modes: 61, 93

4. 101, 194, 121, 153, 101

   mean: 134; median: 121; mode: 101

5. 4.8, 5.7, 2.1, 2.1, 4.8

   mean: 3.9; median: 4.8; modes: 2.1, 4.8

The table below shows the average monthly temperatures in °F in Atlanta, GA, for 1996. Use the table to answer Exercises 6–9.

Month	Temperature	Month	Temperature
Jan	54	Jul	88
Feb	57	Aug	88
Mar	63	Sep	83
Apr	72	Oct	74
May	81	Nov	62
Jun	87	Dec	53

6. Find the mean temperature. **71.8°F**

7. Find the median temperature. **73°F**

8. Find the mode for the average monthly temperatures. **88°F**

9. Which measure best describes the average monthly temperatures? Why?

   Mean or median; both are central values.

Mixed Applications

10. Jenny's math test scores for the semester are 96, 88, 76, 95, 100, and 85. What are the mean, median, and mode for Jenny's tests?

    90; 91.5; no mode

11. The mean of Mr. Moore's salary for the past 3 years is $26,500. If he earned $23,000 his first year and $26,000 his second year, how much did he earn his third year?

    $30,500

12. The difference between two numbers is 23.7. Their sum is 61.1. What are the two numbers?

    42.4 and 18.7

13. Marcus is serving punch at his birthday party. He divides it equally among 12 guests. Each guest drinks $2\frac{3}{4}$ c of punch. How many cups of punch does Marcus serve?

    33 cups

Name ______________________

LESSON 21.3

Using Appropriate Graphs

Vocabulary

Complete.

1. Two graphs that show the relationship between the parts and the whole are a circle graph and a(n) **stacked bar graph**.

---

2. Make an appropriate graph with the given data. Choose between a line graph and a bar graph. Explain your choice of graph.
Check students' graphs.
bar graph; categorical data

Favorite Colors of Second Graders					
red	blue	green	black	orange	purple
14	24	9	12	5	6

3. Sketch a stacked bar graph for the given data. **Check students' graphs.**

Household Expenses	
Food	22%
Clothing	7%
Housing	35%
Transportation	14%
Other	22%

For Exercises 4–5, use the box-and-whisker graph at the right, which compares the test scores of two seventh-grade math classes.

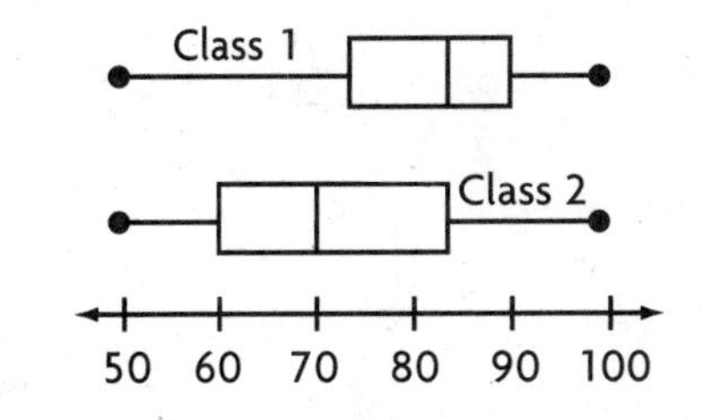

4. What do the whiskers show?
extreme values and spread of scores

5. Which range is greater? Which median is less?
They are the same; Class 2.

Mixed Applications

6. Suppose you have a bag of 100 candies. You have green, orange, red, tan, brown, and yellow candies. What percent of each color would you like? Make a graph in the space at the right.

Check students' graphs.

7. A radio talk-show host asks the name of every tenth caller's political party. What type of sample does he use?
systematic sample

Name ______________________

Misleading Graphs

For Exercises 1–3, use the graph at the right.

1. How many books are mysteries? science fiction?

   2,000; 10,000

2. What is the ratio of science fiction books to mystery books? What ratio does the graph seem to show?

   5 to 1; about 3 to 1

3. How can you change the graph so that it is not misleading? Change the scale.

For Exercises 4–6, use the graph at the right.

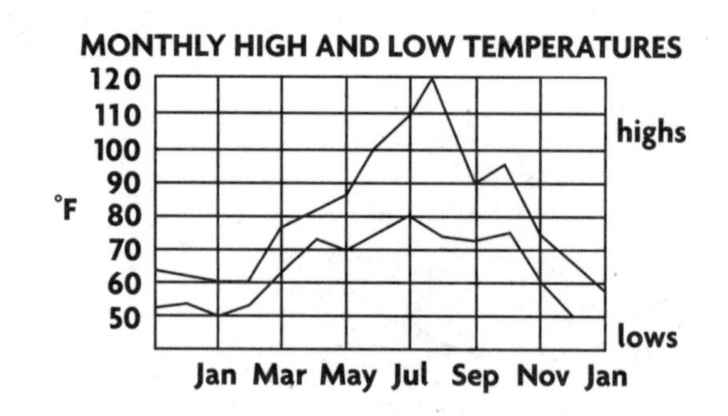

4. What was the July high temperature? the July low temperature?

   110°F; 80°F

5. What is the ratio of this high temperature to this low temperature? What ratio does the graph seem to show?

   11 to 8; about 3 to 1

6. How can you change the graph so it is not misleading? Change the scale.

Mixed Applications

For Problems 7–8, use the graph.

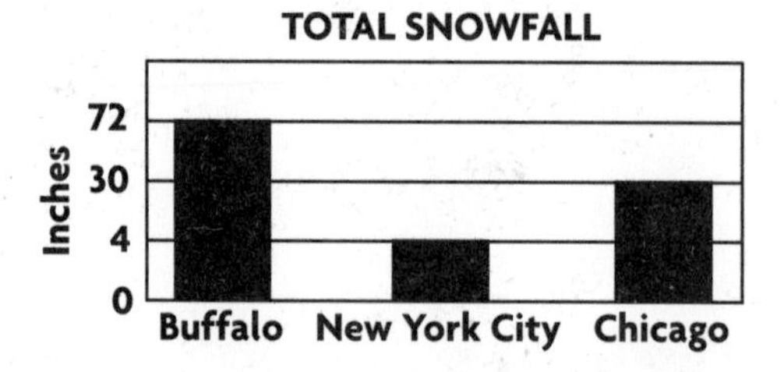

7. What is the approximate ratio of snowfall in Chicago to snowfall in New York City? What ratio does the graph seem to show?

   about 8 to 1; 2 to 1

8. How could the graph be corrected?

   Change the scale.

9. Andrew scored 75, 67, 82, and 86 on four history tests. If he wants at least an 80 average, what must he score on his fifth test?

   at least 90

10. Circus acrobats are making a human pyramid 5 levels high. How many people will they need?

   15 people

Use with text pages 410–411.

Name ______________________

LESSON 22.1

Tree Diagrams and Sample Spaces

Vocabulary

Complete.

1. The set of all possible outcomes is called the ___sample space___.

2. The ___Fundamental Counting Principle___ states that you can multiply the number of different ways each choice can occur to find the total number of possible outcomes.

For Exercises 3–5, use the information at the right. Make a tree diagram of the sample space. Tell the total number of outcomes. Check students' tree diagrams.

3. The "Full Dinner" allows you to choose an entree and a side order.

   ___12 outcomes___

4. The "Dinner for 1" allows you to choose a soup and an entree.

   ___16 outcomes___

5. The "Dinner Special" allows you to choose an appetizer and an entree.

   ___20 outcomes___

Chou's Chinese Kitchen

Soups:	Appetizers:
Egg Drop	Egg Roll
Won Ton	Bar-B-Q Pork
Hot and Sour	Pot Stickers
Chicken Noodle	Fried Wonton
	Crab Rangoon
Entrees:	**Side Orders:**
Beef with Vegetables	Rice
Shrimp with Broccoli	Pan-fried Noodles
Chicken with Pea Pods	Crispy Noodles
Pork Almond Ding	

Mixed Applications

6. At the zoo, a giraffe 3.2 m tall is standing by a goat. The giraffe's shadow is 1.8 m and the goat's shadow is 0.9 m. How tall is the goat?

   ___1.6 m___

7. Billy has 4 shirts, 3 pants, and 2 sweaters. How many three-piece outfits are possible? Make a tree diagram of the sample space. Then solve. Check students' tree diagrams.

   ___24 outfits___

8. Mr. Tamanaha earns a 15% commission on everything he sells. If he sells 4 CD players for $99 each and 2 keyboards for $149 each, how much does he earn?

   ___$104.10___

Name ______________________

LESSON 22.2

Finding Probability

Vocabulary

Complete.

1. The number used to describe the chance of an event's occurring is called the mathematical probability.

A number cube is labeled with the numbers 1, 3, 5, 7, 9, and 11. Find the probability.

2. P(3, 5, or 7) $\frac{1}{2}$
3. P(odd number) 1
4. P(even number) 0
5. P(number greater than 7) $\frac{1}{3}$
6. P(factor of 45) $\frac{2}{3}$
7. P(multiple of 3) $\frac{1}{3}$
8. P(number less than 1) 0
9. P(1 or 3) $\frac{1}{3}$
10. P(prime number) $\frac{2}{3}$
11. P(square root of 9) $\frac{1}{6}$
12. P(factor of 12) $\frac{1}{6}$
13. P(multiple of 5) $\frac{1}{6}$

When the probability of an event is close to 0, the event is unlikely to happen. When the probability is close to 1, the event is likely to happen. For Exercises 14–15, find the probability. Then tell whether the event is likely or unlikely to happen.

14. What is the probability that Paul will spin a 2 on a spinner with 8 equal sections labeled 1–8?

$\frac{1}{8}$; unlikely

15. What is the probability that Myrna will pick a striped marble from a bag containing 3 blue marbles and 12 striped marbles?

$\frac{4}{5}$; likely

Mixed Applications

16. Scott's bank contains 3 pennies, 5 nickels, 4 dimes, and 8 quarters. Scott shakes the bank and one coin falls out. What is the probability that it is a dime?

$\frac{1}{5}$

17. Joseph's family is flying to Puerto Rico to visit relatives. The distance they will fly is 2,250 mi. They know the trip will take 3 hr. What is the rate of speed of the plane, in miles per hour?

750 mph

18. A bag of candy contains 10 yellow, 5 red, 8 green, 7 orange, and 16 brown candies. What is the probability of selecting a green or brown candy?

$\frac{12}{23}$

19. A pizza parlor offers pizzas in 3 sizes, with a choice of 3 crusts and 6 toppings. How many different pizza choices are possible?

54 pizza choices

Use with text pages 418–419.

Name ______________________________

Problem-Solving Strategy

Make a List: Combinations and Probability

Make a list and solve.

1. Farmer McDonald has 6 pigs to feed, but room at the feeding trough for only 2 at a time. How many combinations of 2 pigs at the trough can be made from 6 pigs?

   15 combinations

2. There are 5 empty seats in Shawna's Cafe. Monya and 6 friends want to eat there. How many combinations of 5 students can be made from 7 students?

   21 combinations

3. Yolanda's Yogurt Shop offers sundaes with a choice of 3 of 6 toppings: nuts, sprinkles, whipped cream, hot fudge, cherries, and coconut. How many combinations of 3 toppings can be made from 6 choices? What is the probability that a sundae will not have whipped cream and hot fudge together?

   20 different combinations; $\frac{1}{5}$

4. Jiffy Foods is interviewing 8 people for 2 openings. How many combinations of 2 people can be hired from the 8 people being interviewed?

   28 combinations

Mixed Applications

Solve.

CHOOSE A STRATEGY

- Make a List
- Use a Formula
- Work Backward
- Draw a Diagram

Choices of strategies will vary.

5. Donald buys 5 different packets of flower seeds. He decides to plant only 3 of them. How many combinations of 3 packets can be made from 5 seed packets? Two of the flower seeds are daisy and columbine. What is the probability that daisy and columbine are among the 3 packets?

   10 combinations; $\frac{3}{10}$

6. Lucinda has onions, tomatoes, cabbage, potatoes, carrots, celery, and peas in the refrigerator. She uses 4 of the vegetables in a stew. How many combinations of 4 vegetables can be made from 7 vegetables?

   35 combinations

7. If all the digits are different, how many 3-digit odd numbers greater than 700 can be written using the digits 1, 2, 3, 5, 6, and 7?

   12 numbers

8. Helga paid for her groceries with a check. She bought a ham for $7, two bags of oranges for $4.98 each, and a gallon of milk for $2.29. She got $15 back. What was the amount of the check?

   $34.25

Use with text pages 421–423.

Name ________________________________

LESSON 22.4

Finding Permutations and Probability

Vocabulary

Complete.

1. A ___permutation___ of items or events is an arrangement in which the order is important.

Natasha writes the letters M, A, T, and H, one to an index card. She wants to see how many different ways she can arrange the 4 cards.

2. How many permutations will have the letter "M" first?

   ___6 permutations___

3. How many permutations will have "M" first and "T" second?

   ___2 permutations___

Write a multiplication equation to find the number of permutations.

4. Joy is taking math, science, English, history, and band. Find the number of possible orders for the 5 classes.

   $5 \times 4 \times 3 \times 2 \times 1 = 120$

5. Frasier's Bookstore has 6 best-sellers they want to display in their window. Find the number of possible orders in which they can display the 6 books.

   $6 \times 5 \times 4 \times 3 \times 2 \times 1 = 720$

Find the probability.

6. You have 5 cards in a box, each labeled with one letter of the word MAGIC. What is the probability that you will draw the letters MAGIC in order?

   $\frac{1}{120}$

7. Joseph and Abe are playing a game. Abe rolls two number cubes, each numbered from 1 to 6. What is the probability of rolling two 6's?

   $\frac{1}{36}$

Mixed Applications

8. At the school track meet, there are 8 runners in the 100-yd dash. If there are no ties, in how many different ways can first, second, and third prizes be given?

   $8 \times 7 \times 6 = 336$ ways

9. Alexa, Steve, Henry, Anna, Jan, and Juan go to a movie. They find 6 empty seats at the end of a row. What is the probability that the friends sit in alphabetical order?

   $\frac{1}{720}$

10. Dexter has $\frac{1}{5}$ the cost of a CD. How much does the CD cost if Dexter has \$2.80? Write an equation for the word sentence, and then solve. **Variables may vary.**

    $\frac{1}{5}c = 2.80$; $c = 14$; cost = \$14

11. A sandbox has a perimeter of 24 ft and an area of 35 ft^2. What are the dimensions of the sandbox?

    $5 \text{ ft} \times 7 \text{ ft}$

Use with text pages 424–427.

Name ______________________

LESSON 23.1

Experimental Probability

Vocabulary

Complete.

1. The **experimental probability** of an event is the ratio of the number of times the event occurs to the total number of trials, or times you do the activity.

The game piece shown at the right has just four sides, labeled *a, b, c,* and *d.* In Exercises 2–3, each table shows the results of rolling the game piece 50 times. Find the probability of rolling each letter.

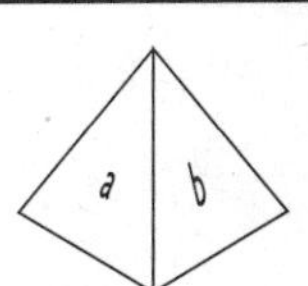

2.

Letter	*a*	*b*	*c*	*d*
Times rolled	15	9	8	18

a: $\frac{3}{10}$; b: $\frac{9}{50}$; c: $\frac{4}{25}$; d: $\frac{9}{25}$

3.

Letter	*a*	*b*	*c*	*d*
Times rolled	11	14	17	8

a: $\frac{11}{50}$; b: $\frac{7}{25}$; c: $\frac{17}{50}$; d: $\frac{4}{25}$

4. Rick flips a coin 50 times and gets 10 heads. Why is this surprising?

You would expect 25 of 50 flips to be heads.

5. The spinner at the right has five spaces of equal size. What is the mathematical probability of the pointer stopping in the space marked 5? If you spin the pointer 30 times, how many times would you expect it to stop on 5?

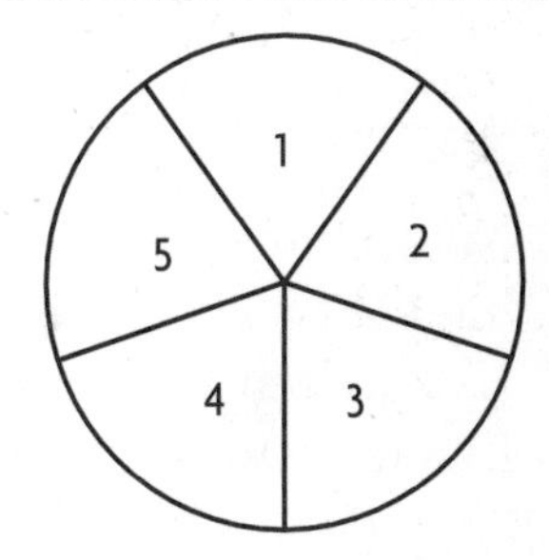

$\frac{1}{5}$; 6 times

Mixed Applications

6. A baseball player hit the ball 42 times out of the last 120 times at bat. What is the player's probability of getting a hit the next time at bat?

$\frac{42}{120}$, or $\frac{7}{20}$

7. Max makes handmade doll houses, which he sells for $55 each. This is 4 times his cost, plus $7. What is Max's cost?

$12

8. For the first half of the season, a baseball player hit the ball 17 times out of 54 at-bats. During the second half, the player had 19 hits for the same number of at-bats. What is the probability that she will get a hit in her next official at-bat?

$\frac{1}{3}$

9. The trombone section of the school band has 6 trombonists. The band director will select 2 trombonists to play in the town band. How many combinations of 2 musicians can be made from 6 musicians?

15 combinations

Name ______________________________

Problem-Solving Strategy

Acting It Out by Using Random Numbers

A computer-generated list of random numbers 1–8 is shown at the right. Use the list for Exercises 1–3.

For Ex. 1–5, possible answers are given for a left corner, top row starting point and a movement from left to right.

2	5	1	8	3	2	1
7	1	8	2	5	3	4
3	4	6	1	2	8	2
6	3	2	7	2	8	4

1. You are playing a game with a spinner that has 8 congruent sections labeled 1–8. How many spins will it take before you get all the numbers 1 through 8?

   17 spins

2. Sparkle Cereal has a winning game piece in 2 of every 8 boxes of cereal. Let the numbers 4 and 7 represent winning game pieces. How many boxes of cereal would you have to buy to win?

   8 boxes of cereal

3. A student randomly guesses the answers to 14 true-false questions. How many questions is the student going to answer as *false*? Let an even number mean *true* and an odd number mean *false*.

   8 questions

Mixed Applications

Solve.

CHOOSE A STRATEGY

- Draw a Diagram • Act It Out • Find a Pattern • Use a Formula • Write an Equation • Make a Table

Choices of strategies will vary.

4. Natasha put 8 cards numbered 1–8 in a bag. Ted picks a card and puts it back. Use the random number table above to find how many times Ted will pick the 2 card if he picks 25 times.

   6 times

5. At a basketball game, 2 of every 8 tickets have a coupon for a free soda. Let the numbers 1 and 4 represent winning tickets. Use the random number table above to find how many tickets you would have to buy to win a soda.

   3 tickets

6. At 2 P.M. the air temperature was 18°F. This was 34° higher than at 2 A.M. What was the temperature at 2 A.M.?

   ⁻16°F

7. Vince paid $9 more for a blanket than for sheets. He paid $45 for both. How much did he pay for the blanket?

   $27

8. Ron is putting a fence around a rectangular garden. The length of the garden is three times the width. If the garden is 6 ft wide, how much fencing does Ron need?

   48 ft

9. You enter a hotel elevator. You go up 8 floors, down 2 floors, and up 6 floors. If you entered the elevator on the fifth floor, what floor are you on now?

   17th floor

Use with text pages 436–437.

Name ______________________

Designing a Simulation

Vocabulary

Complete.

1. A **simulation** is a model of an experiment that would be too difficult or time-consuming to actually perform.

For Problems 2–5, design a simulation in which a spinner, number cube, or coin is used to model each situation. **Possible answers are given.**

2. A soccer goalie who successfully rejects 50% of the shots on goal is set for the final shot of the game.

   coin, with heads meaning the goalie rejects the shot

3. A field-goal kicker who makes 2 out of 3 field-goal attempts is kicking a field goal to win the game.

   a spinner with 3 equal sections labeled 1, 2, and 3, odd-numbered sections meaning the field goal is good

4. One of every three grapes is sour.

   number cube with rolls of 1 or 2 meaning "sour"

5. Each day the lunch supervisor selects 1 table out of 8 tables to begin the lunch line.

   spinner with 8 equal sections

Mixed Applications

6. A bakery randomly puts a coupon for a free loaf in 1 of every 4 packages of bread. Design and conduct a simulation to model finding a coupon in a package.

   Possible answer: Use a spinner with 4 equal sections, with section 1 meaning "free loaf coupon."

7. Last month, 2 out of every 12 customers of the Sweat Shoppe bought a treadmill. The owner estimates that about 8 of his first 41 customers this month will buy a treadmill. Design a simulation to check his estimate.

   Possible answer: Use a number cube with 1 meaning "treadmill."

8. There are 32 pens in a box, 4 red, 6 green, 12 blue, and 10 black. What is the probability of randomly picking a blue pen? a red pen? a green pen?

   $\frac{3}{8}, \frac{1}{8}, \frac{3}{16}$

9. Tia sells homemade cheesecakes. She sells a cheesecake for 5 times her cost, plus $5. The price of a cheesecake is $22.50. What is her cost?

   $3.50

Name ______________________________

Geometric Probability

Vocabulary

Use *a, b,* or *c* to complete.

1. A probability you calculate by comparing the area of a specific region to the area of the larger region it is in is called a(n) __b__ probability.

   a. experimental b. geometric c. mathematical

Find the probability that a dart that hits the target will randomly land in the shaded area. All measurements are in inches.

2. 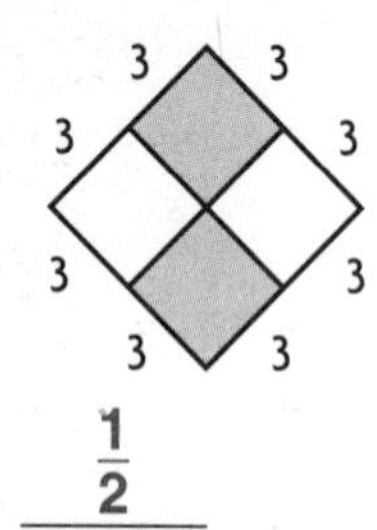

$\frac{1}{2}$

3. 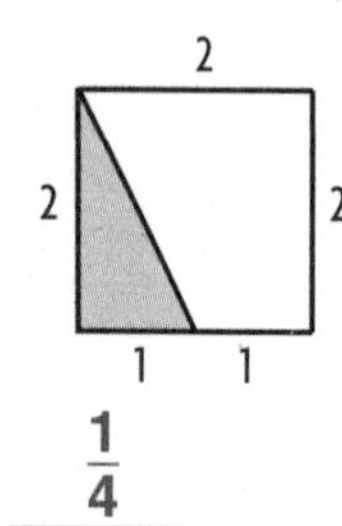

$\frac{1}{4}$

4. 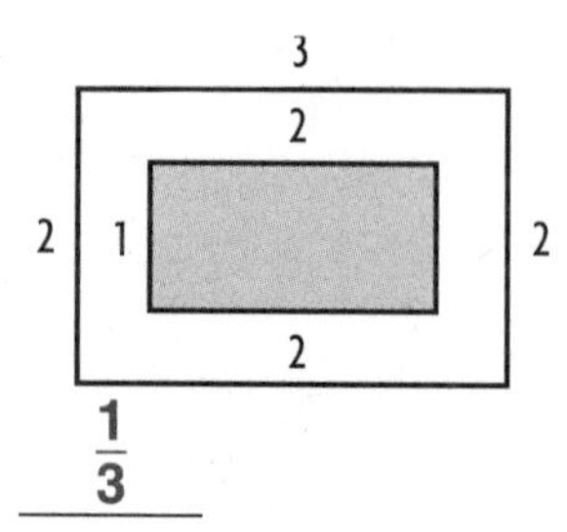

$\frac{1}{3}$

Suppose a meteor lands somewhere at random in one of the oceans, which have a total area of 129,397,000 mi^2. Find the probability that the meteor will land in the given ocean. Give your answers as percents to the nearest tenth.

5. Pacific Ocean (about 63,855,000 mi^2)

   about 49.3%

6. Atlantic Ocean (about 31,744,000 mi^2)

   about 24.5%

7. Indian Ocean (about 28,371,000 mi^2)

   about 21.9%

8. Arctic Ocean (about 5,427,000 mi^2)

   about 4.2%

Mixed Applications

9. At Taylor's birthday party, children are playing "Pin the Tail on the Donkey." If the paper donkey has an area of 28 $in.^2$, what is the probability that the pin for the tail will be placed inside a circle with a diameter of 2 in.?

   $\frac{\pi}{28}$, or about 11.2%

10. Friendship Park measures 850 ft by 1,200 ft. At the center of the park is a flower garden that measures 136 ft by 60 ft. What is the probability that a lightning bolt that hits the park will strike the garden?

    0.8%

11. Leon picked 39 tomatoes from his garden. This was 30% of all the tomatoes. How many tomatoes did he grow?

    130 tomatoes

12. In how many ways can a soccer coach choose the goalie, left wing, midfielder, and right wing from a team of 8 players?

    1,680 ways

Use with text pages 441–443.

Name ______________________

Measuring and Estimating Lengths

Vocabulary

Write the correct letter from column 2.

1. precision ___b___
2. greatest possible error ___a___

a. $\frac{1}{2}$ of the unit used in a measurement
b. the smallest unit of measurement used

Give the precision of each measurement.

3. 10 yd ___1 yd___
4. 5 cm ___1 cm___
5. $6\frac{1}{4}$ ft ___$\frac{1}{4}$ ft___
6. $2\frac{1}{2}$ in. ___$\frac{1}{2}$ in.___

Find the greatest possible error (GPE) for each measurement.

7. 13 km ___$\frac{1}{2}$ km___
8. $4\frac{1}{2}$ in. ___$\frac{1}{4}$ in.___
9. 15 ft ___$\frac{1}{2}$ ft___
10. 2.1 m ___0.05 m___

For the given measurement, give the smallest and largest possible actual lengths.

11. 16 cm
___$15\frac{1}{2}$ cm; $16\frac{1}{2}$ cm___
12. $12\frac{1}{2}$ ft
___$12\frac{1}{4}$ ft; $12\frac{3}{4}$ ft___
13. 9 km
___$8\frac{1}{2}$ km; $9\frac{1}{2}$ km___
14. 14.2 in.
___14.15 in.; 14.25 in.___

15. Would you estimate the length of a math book as 1 ft or 1 m? ___1 ft___

16. Would you estimate the length of a pencil as 18 cm or 18 in.? ___18 cm___

17. Would you estimate the length of a credit card as 6 cm or 6 dm? ___6 cm___

Mixed Applications

18. After a thunderstorm, Sue found a worm 5 in. long. What is the greatest possible error of the measurement?

___$\frac{1}{2}$ in.___

19. If there are 15 boys in a class of 35 boys and girls, what is the probability that the president of the class is a girl?

___$\frac{4}{7}$___

20. Nancy measured the length of her desk as 1 m. Brad measured the same length as 100 cm. Which measurement is more precise? Explain.

___100 cm; the smaller the unit of measure, the more precise the measurement.___

21. Drew drops a rubber ball from a height of 32 ft. After each bounce, the ball rises to $\frac{1}{2}$ its previous height. After how many bounces is the ball's highest point 2 ft off the ground?

___4 bounces___

Use with text pages 452–455.

Name ____________________

Networks

Vocabulary

Select the correct terms to complete the definition.

a. edges **b.** graph **c.** vertices

1. A network is a(n) __b__ with __c (or a)__ and __a (or c)__.

Starting from *A*, find all the possible routes that include all the cities. Each vertex represents a city.

2. *ABCDEF,* *ABCFED,* *AFEDCB*

3. *ABCDEF, AFDECB,* *AFEDCB, ABCEDF,* *ABCEFD*

Starting from A, find all the possible routes that include every vertex. Find the distance for each. Distances are in kilometers.

4.

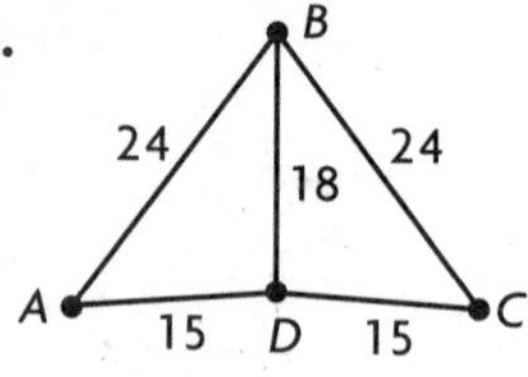

ADBC, 57 km
ABCD, 63 km
ADCB, 54 km
ABDC, 57 km

5.

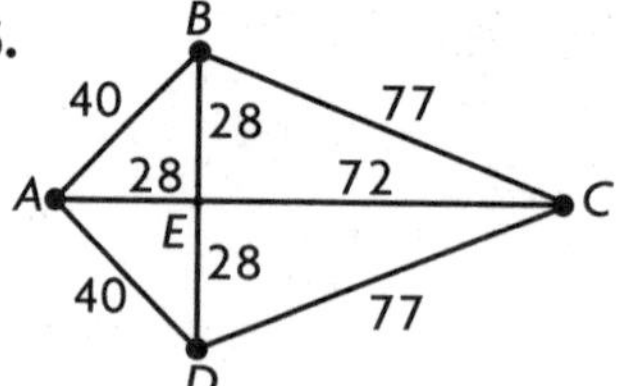

ABCDE, 222 km
ABCED, 217 km
ADCBE, 222 km
ADCEB, 217 km
AEBCD, 210 km
AEDCB, 210 km

6. In Exercise 4, what is the shortest route? the longest route?

ADCB = 54 km; *ABCD* = 63 km

Mixed Applications

7. The diagram shows Maria's office computer network. Which computers are not directly connected?

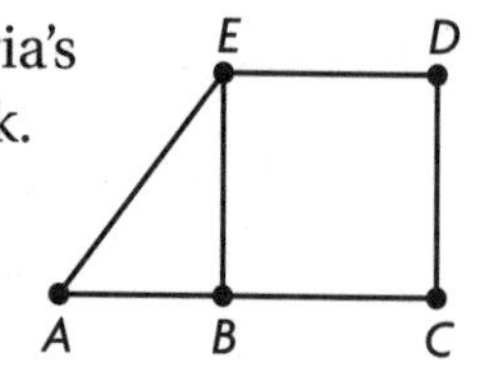

A and *C*, *A* and *D*, *B* and *D*, *C* and *E*

8. Matthew wants to buy a sports car. It is available in 6 exterior colors and 4 interior colors. From how many color combinations can Matthew choose?

24 color combinations

Use with text pages 456–457.

Name ______________________

LESSON 24.3

Pythagorean Property

Vocabulary

For Exercise 1, complete the statement. For Exercise 2, fill in the blanks.

1. In a right triangle with sides of lengths of a, b, and c, the relationship

   $a^2 + b^2 = c^2$ is the **Pythagorean Property**.

2. sides adjacent to right angle: **legs** side opposite right angle: **hypotenuse**

Name the legs and the hypotenuse of each right triangle.

3. 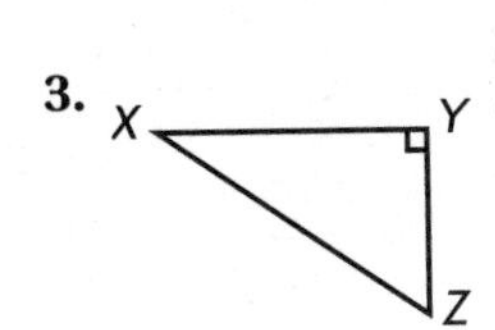

   $\overline{XY}$ and $\overline{YZ}$ are legs,

   $\overline{XZ}$ is the hypotenuse.

4. 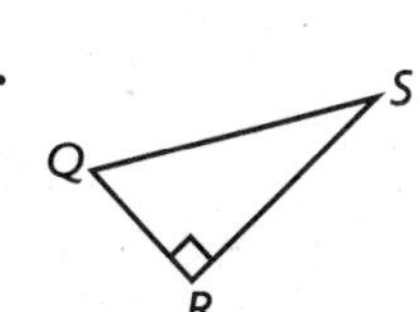

   $\overline{RS}$ and $\overline{QR}$ are legs,

   $\overline{QS}$ is the hypotenuse.

Tell whether the three sides form a right triangle. Write *yes* or *no*.

5. 12 in., 16 in., 20 in. **yes**
6. 30 mi, 40 mi, 50 mi **yes**
7. 6 ft, 11 ft, 13 ft **no**
8. 2 cm, 3 cm, 5 cm **no**
9. 5 mm, 12 mm, 13 mm **yes**
10. 8 cm, 15 cm, 17 cm **yes**

Find the length of the hypotenuse for each right triangle. Round to the nearest tenth when necessary.

11. 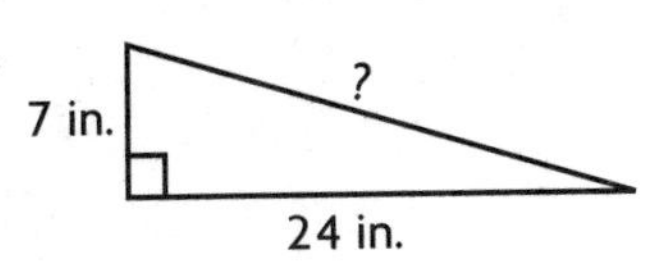

    25 in.

12. 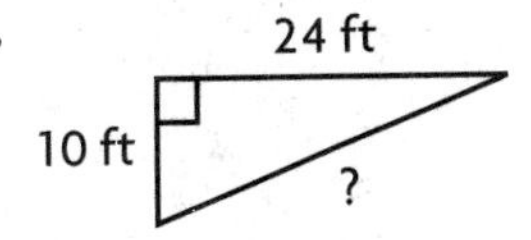

    26 ft

13. 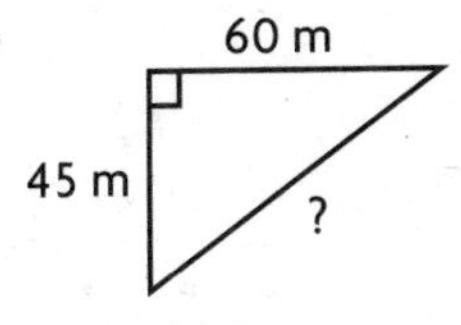

    75 m

14. $a = 8$ ft
    $b = 15$ ft
    $c =$ **17 ft**

15. $a = 3.5$ m
    $b = 12$ m
    $c =$ **12.5 m**

16. $a = 30$ in.
    $b = 72$ in.
    $c =$ **78 in.**

Mixed Applications

17. June and Roy left the riding stable at 10 A.M. June trotted her horse south at 4 km per hr. Roy galloped his horse east at $7\frac{1}{2}$ km per hr. How far apart were they at noon?

    17 km

18. Determine if the triangles are congruent by *SSS*, *SAS*, or *ASA*.

    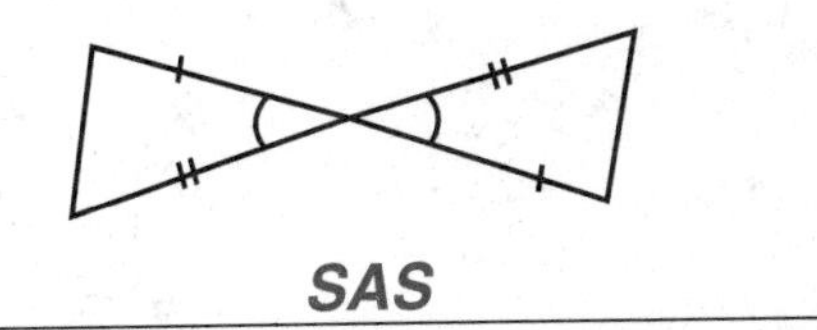

    SAS

Name ______________________

LESSON 24.4

Problem-Solving Strategy

Using a Formula to Find the Area

1. Trina buys a pizza in the shape of a circle with a diameter of 14 in. What is the area of the pizza?

   153.86 in.^{2}

2. The face of an Egyptian pyramid is triangular. The base measures 210 m and the height of the face measures 90 m. What is the area of the face?

   9,450 m^2

3. The American flag at Camp Resolute is 6 ft long and has an area of 18 ft^2. How many feet wide is the flag?

   3 ft wide

4. Mr. Pelak's lawn sprinkler waters a circular region of lawn with a radius of 10 ft. What is the area watered?

   314 ft^2

Mixed Applications

Solve.

CHOOSE A STRATEGY

- Use a Formula
- Draw a Diagram
- Guess and Check
- Write an Equation

Choices of strategies will vary.

5. Carlos's family wants to carpet his bedroom, which measures 3 yd × 4 yd. The carpet costs $6 per square yard. What will be the cost to carpet Carlos's bedroom?

   $72

6. The sum of two numbers is 91. Half of one number is three times the other number. What are the numbers?

   13 and 78

7. A farm in South Dakota is in the shape of a parallelogram. It is 7 km long and measures 4 km between sides. One half of the land is planted with wheat. What is the area of the land planted with wheat?

   14 km^2

8. The door to Mia's rabbit hutch needs to be braced with a diagonal rod. The door measures 12 in. × 5 in. What is the length of the diagonal brace?

   13 in.

9. Keith bought some candy. He paid $1.80 for taffy and lemon drops. Taffy costs $0.15 per piece and lemon drops cost $0.06 per piece. He bought more taffy than lemon drops. How many of each did Keith buy?

   10 pieces of taffy and 5 lemon drops

10. The Waikiki Yacht Club flag is in the shape of a triangle. The base is 27 in. and the height is 9 in. What is the area of the flag?

    121.5 in.^{2}

Use with text pages 462–464.

Name ______________________

Area of a Trapezoid

Vocabulary

Tell whether the statement is *true* or *false.*

1. A trapezoid has two pairs of parallel sides. __false__

Find the area. Use $A = \frac{1}{2}h(b_1 + b_2)$ for a trapezoid.

2.

7 ft
4 ft
12 ft

__38 ft²__

3.

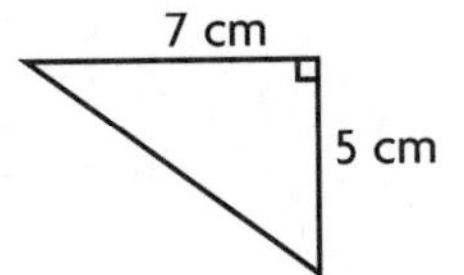

__17.5 cm²__

4.

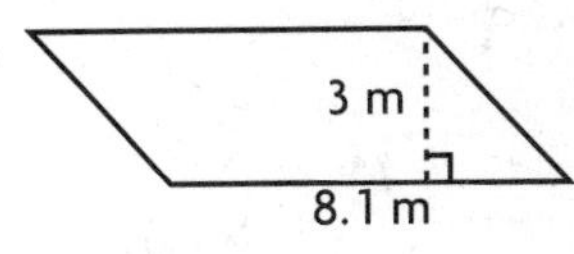

__24.3 m²__

5.

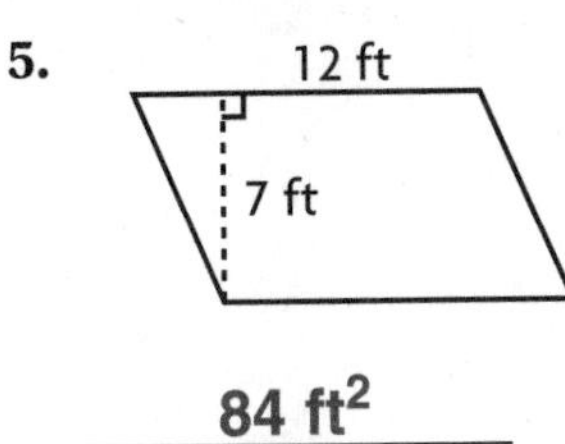

__84 ft²__

6.

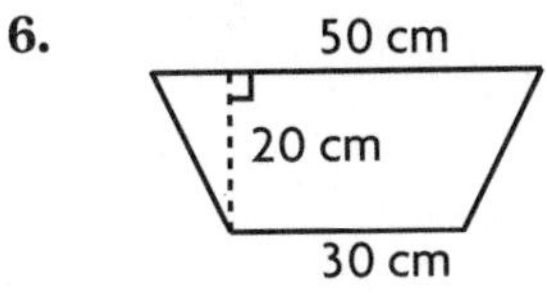

__800 cm²__

7.

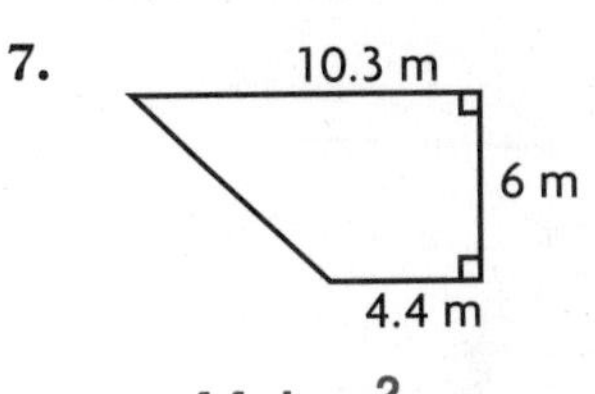

__44.1 m²__

Find the area of the trapezoid.

8. $b_1 = 9$ cm, $b_2 = 24$ cm, $h = 12$ cm — __198 cm²__

9. $b_1 = 12$ ft, $b_2 = 8$ ft, $h = 7$ ft — __70 ft²__

10. $b_1 = 5$ m, $b_2 = 2\frac{1}{2}$ m, $h = 1\frac{1}{4}$ m — __$4\frac{11}{16}$ m²__

11. $b_1 = 5.2$ m, $b_2 = 1.7$ m, $h = 4$ m — __13.8 m²__

12. $b_1 = 4$ in., $b_2 = 7\frac{1}{2}$ in., $h = 2\frac{1}{2}$ in. — __$14\frac{3}{8}$ in.²__

13. $b_1 = 15$ yd, $b_2 = 25$ yd, $h = 16$ yd — __320 yd²__

14. $b_1 = 4.7$ mm, $b_2 = 5.9$ mm, $h = 2$ mm — __10.6 mm²__

15. $b_1 = 2.3$ cm, $b_2 = 4$ cm, $h = 0.8$ cm — __2.52 cm²__

Mixed Applications

16. The sides of a waste basket are in the shape of a trapezoid. Each side has a height of 11 in. and bases that measure 7.5 in. and 9 in. What is the area of the trapezoid?

__90.75 in.²__

17. The Washington County Fair has 10 entries for "Best Heifer." In how many ways can the judges award the first, second, and third prize ribbons?

__720 ways__

Name ______________________

Surface Area of Prisms and Pyramids

Vocabulary

Complete.

1. The sum of the areas of all the surfaces of a solid is called the **surface area**.

Find the surface area of each figure.

2. 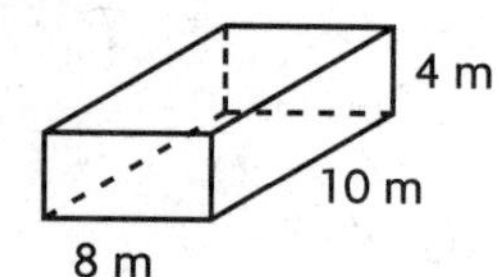

304 m²

3. 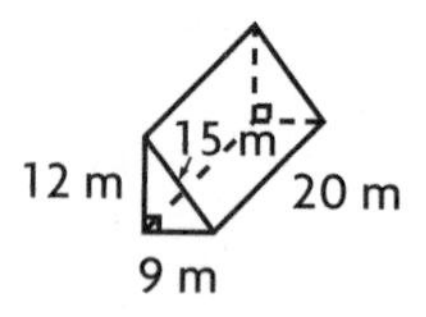

828 m²

4. 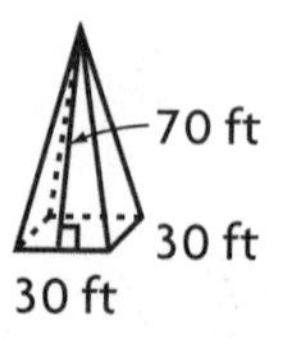

5,100 ft²

5. 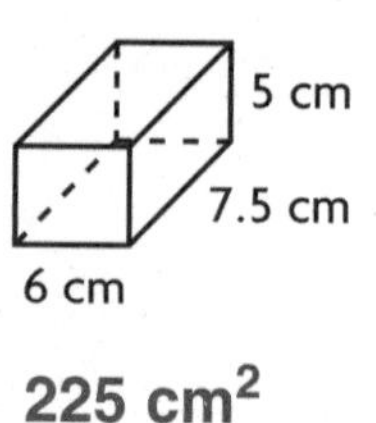

225 cm²

6. 12 m, 9 m, 9 m

297 m²

7. 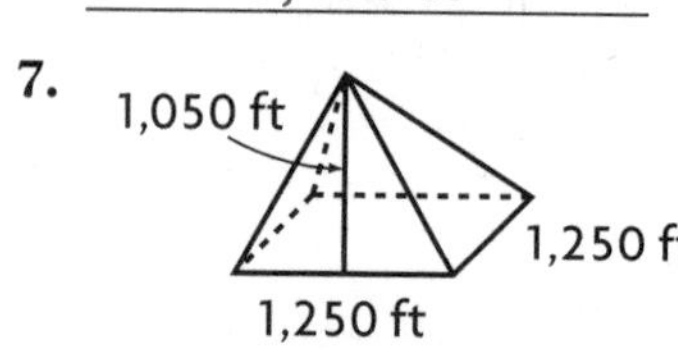

4,187,500 ft²

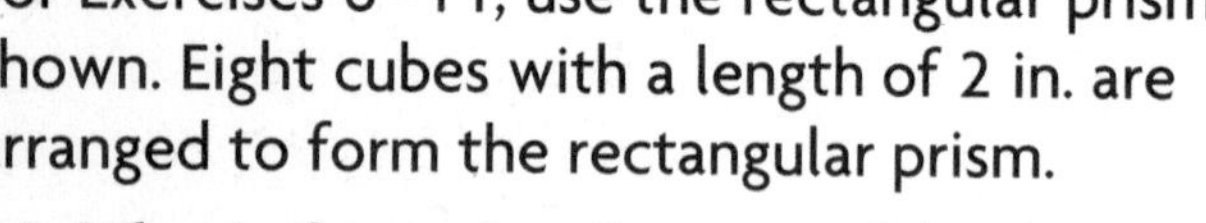
For Exercises 8–11, use the rectangular prism shown. Eight cubes with a length of 2 in. are arranged to form the rectangular prism.

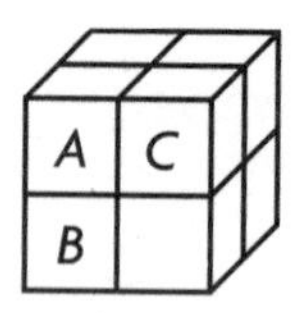

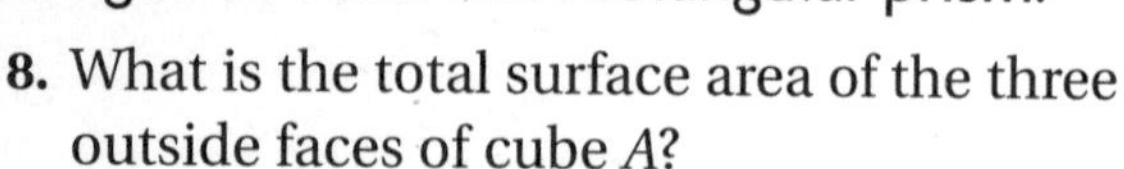
8. What is the total surface area of the three outside faces of cube *A*?

12 in.²

9. What is the total surface area of the rectangular prism?

96 in.²

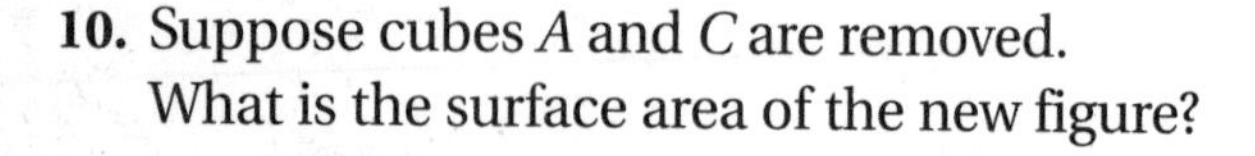
10. Suppose cubes *A* and *C* are removed. What is the surface area of the new figure?

88 in.²

11. Suppose cubes *A* and *B* are removed. What is the surface area of the new figure?

88 in.²

Mixed Applications

12. Miguel is making a model of an Egyptian pyramid. He makes a square pyramid with base 10 in. Each triangular face has a base of 10 in. and a height of 12 in. What is the surface area of his model?

340 in.²

13. The sign in the Museum of Natural History says the wing span of a butterfly in its collection measures $6\frac{3}{4}$ in. What is the greatest possible error of this measurement?

$\frac{1}{8}$ in.

Use with text pages 472–475.

Name ______________________

Finding Surface Area of Cylinders

Vocabulary

1. Write *true* or *false*. The lateral surface is the total surface of a cylinder. **false**

Find the surface area of each figure. Use 3.14 for π. Use $S = 2(\pi r^2) + (2\pi r \cdot h)$.

2.

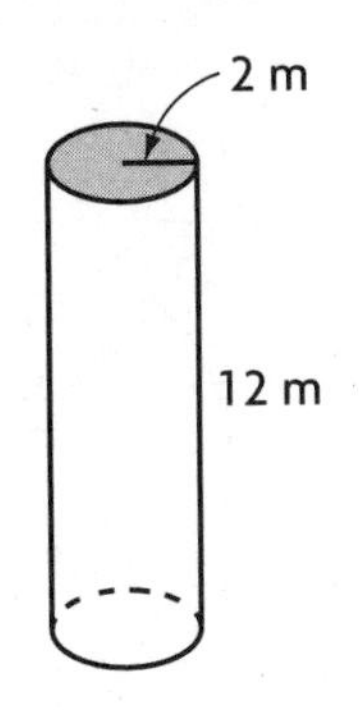

175.8 m²

3.

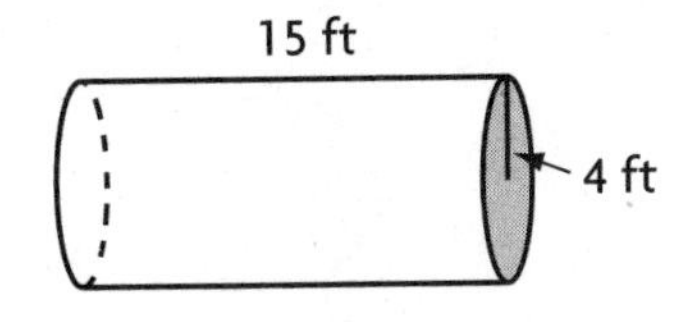

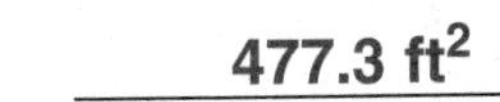

477.3 ft²

4.

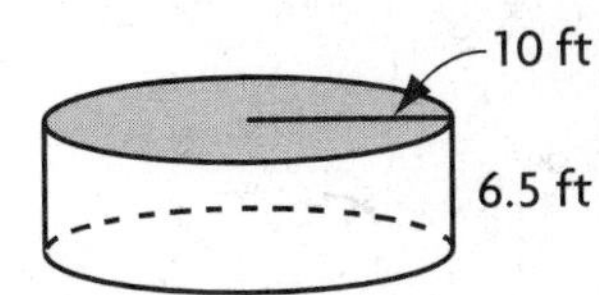

1,036.2 ft²

5.

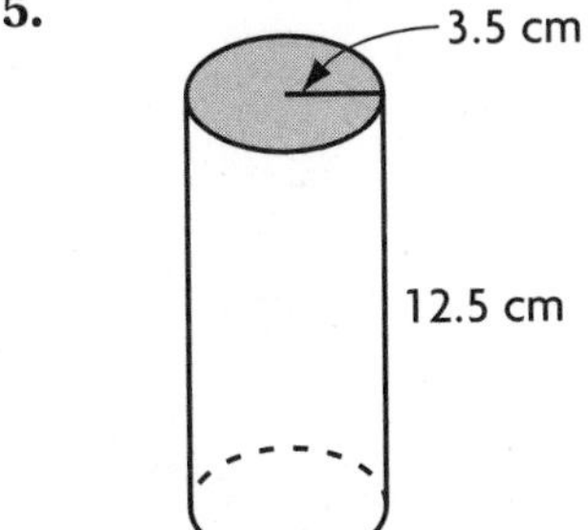

351.7 cm²

6.

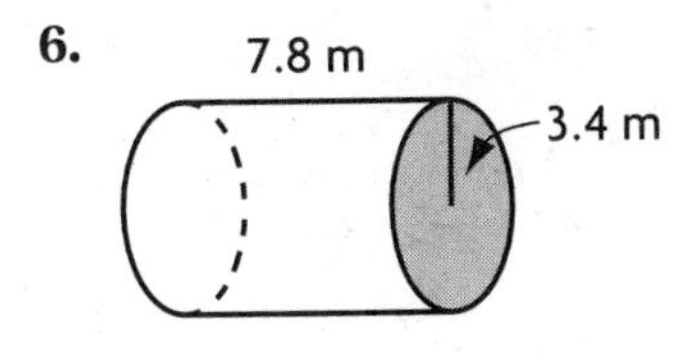

239.1 m²

7.

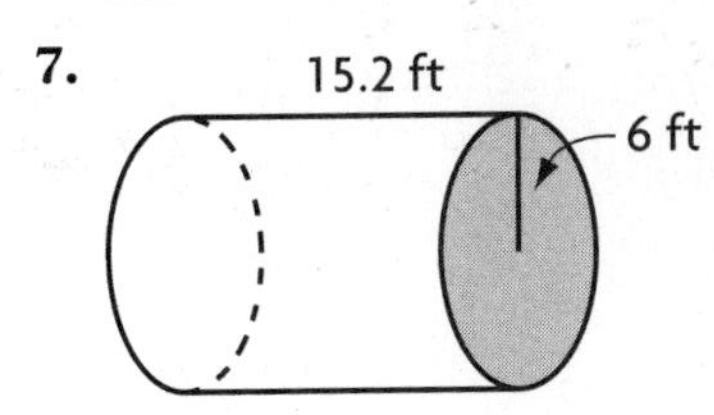

798.8 ft²

Mixed Applications

8. A cylindrical light house is 98 ft high and has a diameter of 12 ft. The lateral surface must be painted. Find the area of the lateral surface.

3,692.6 ft²

9. A group of seventh graders score the following grades: 67, 72, 78, 78, 79, 86, 90, 92, and 98. Find the mean, median, and mode of the scores.

82.2; 79; 78

10. A cylindrical can of frozen grape juice has a diameter of 6 cm and a height of 10 cm. Find the area of the lateral surface. What is the total surface area of the can?

188.4 cm²; 244.92 cm²

11. At 6 P.M. the temperature at International Falls was 46°F. For the next 5 hours the temperature changed at an average rate of $^{-}2.2$°F per hour. What was the temperature at 11 P.M.?

35°F

Use with text pages 476–478.

Name ______________________

LESSON 25.3

Volume of Prisms and Pyramids

Find the volume of each figure. Round your answer to the nearest whole number. Use $V = \frac{1}{3}Bh$ for a pyramid.

1.

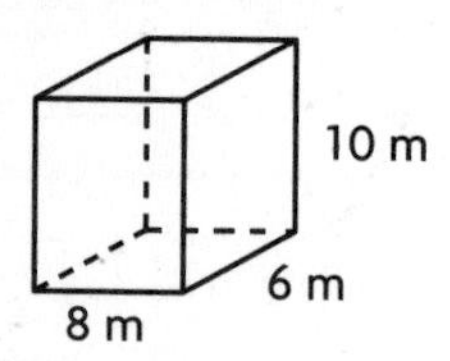

480 m³

2.

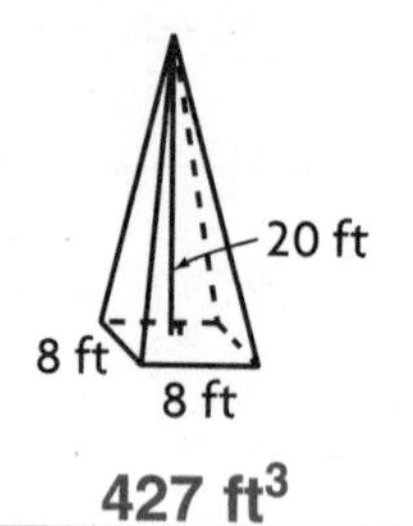

427 ft³

3.

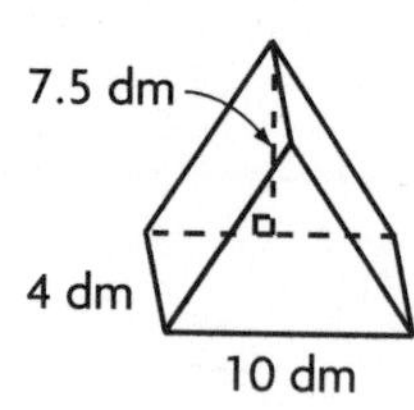

150 dm³

4.

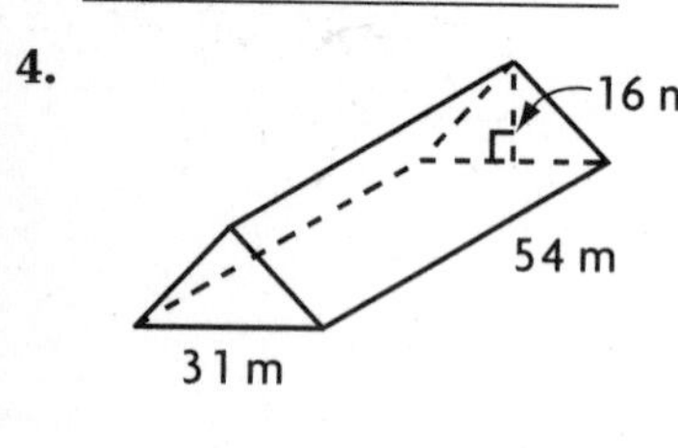

13,392 m³

5.

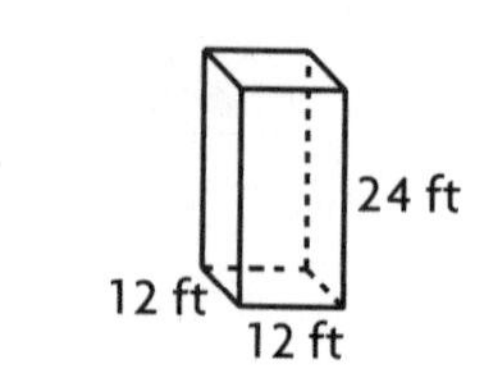

3,456 ft³

6.

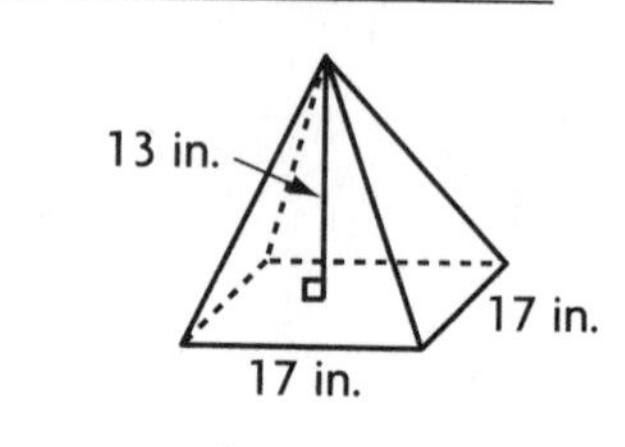

1,252 in.³

For Exercises 7–9, the volume and two dimensions are given for the rectangular prism. Find the missing dimension.

7.

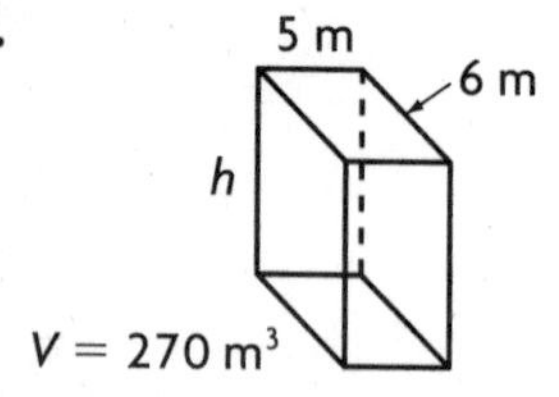

$h = 9$ m

8.

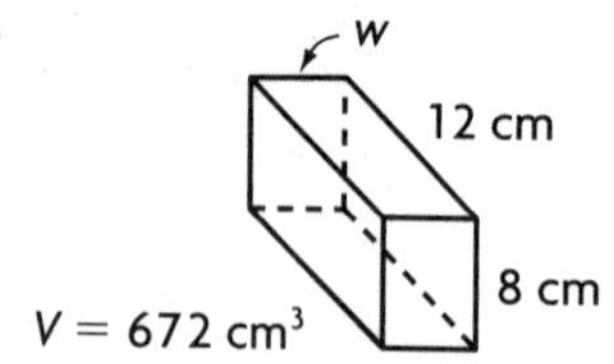

$w = 7$ cm

9.

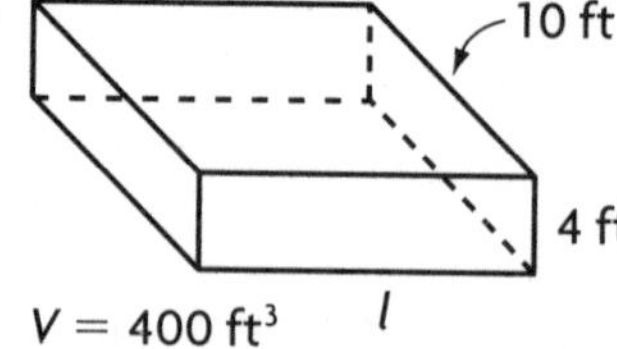

$l = 10$ ft

Mixed Applications

10. An empty office building is 130 ft long, 105 ft wide, and 164 ft high. How many cubic feet of air does the building hold?

2,238,600 ft³

11. A pyramid-shaped tent has a square base that measures 6 ft on each side. The height of the tent is 10 ft. What is the volume of the tent?

120 ft³

12. Ms. Li works for 34 hours at $7.35 per hour. Her employer withholds $68.25 from her paycheck for taxes and insurance. What is the amount of her pay after taxes and insurance?

$181.65

13. Ben has $8.40 in nickels, dimes, and quarters. He has the same number of each. How many of each does he have?

21 nickels, 21 dimes,

and 21 quarters

Use with text pages 480–482.

Name ______________________________

Volume of Cylinders and Cones

Find the volume. Use $\pi = 3.14$, and round to the nearest whole number.
Use $V = \pi r^2 h$ for a cylinder and $V = \frac{1}{3}\pi r^2 h$ for a cone.

1.

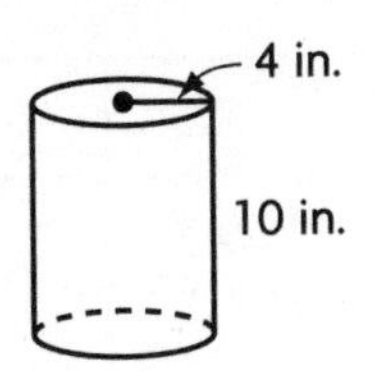

502 in.^{3}

2.

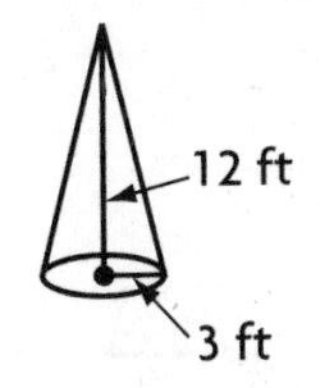

113 ft^3

3.

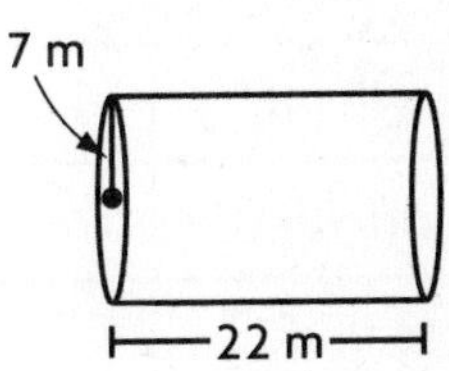

3,384 m^2

4.

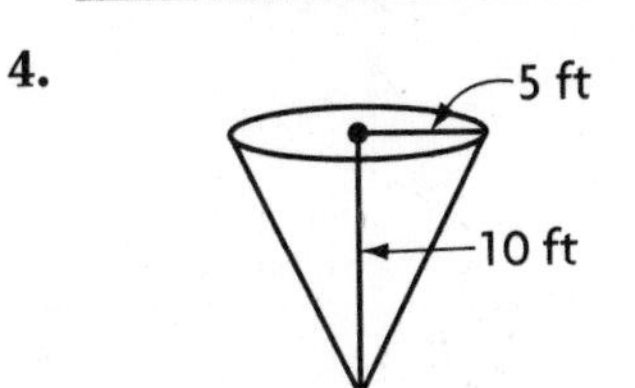

262 ft^3

5.

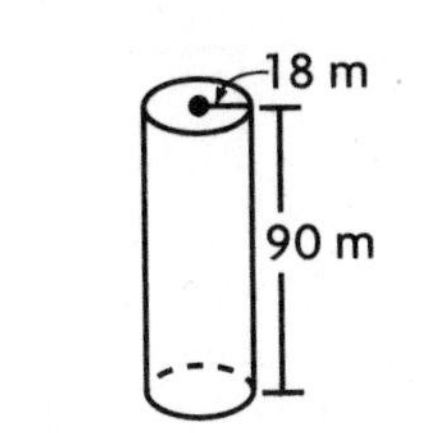

91,562 m^3

6.

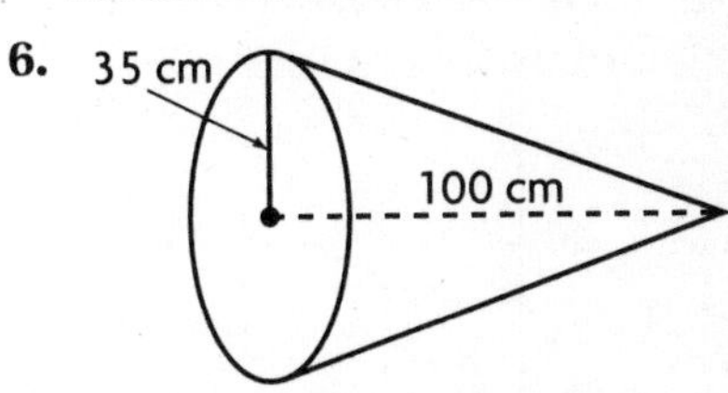

128,217 cm^3

7.

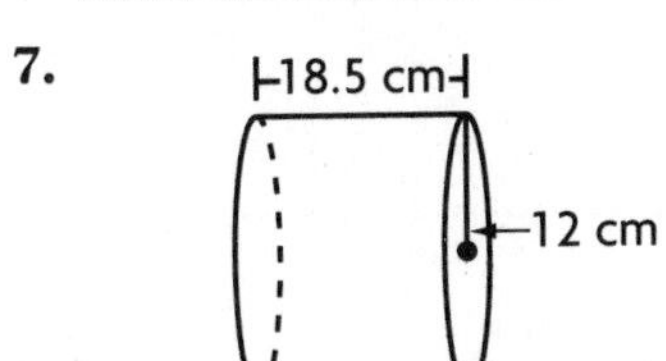

8,365 cm^3

8.

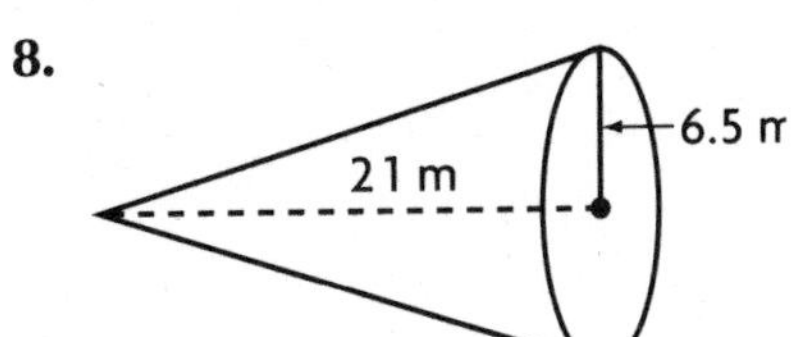

929 m^3

9.

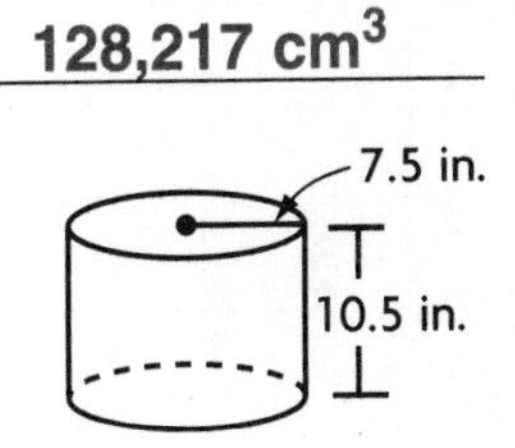

1,855 in.^{3}

10.

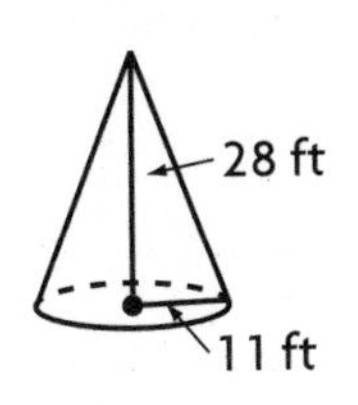

3,546 ft

11.

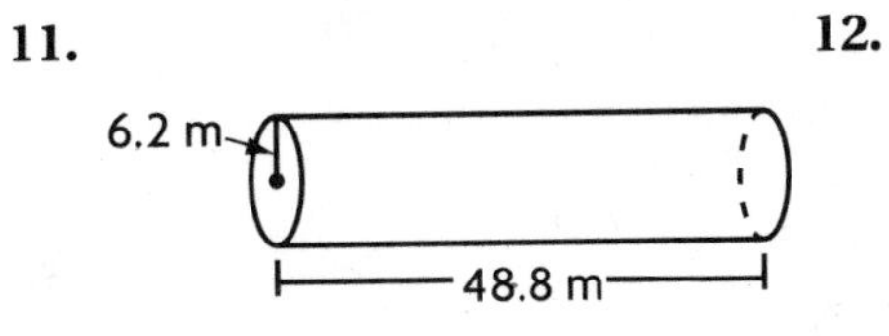

5,890 m^3

12.

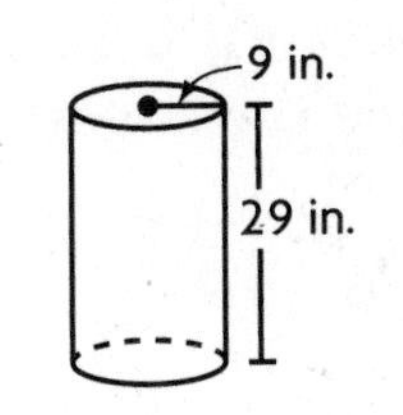

7,376 in.^{3}

Mixed Applications

13. A cone-shaped storage house holds sand for the highway department. If the area of the base is 615 ft^2 and the height is 21 ft, what is its volume?

4,305 ft^3

14. A given plane's distance is given by the formula $d = 500t$. What do you think the variables d and t represent?

d = distance; t = time

15. A cylindrical olive jar is 6 in. high and has a volume of 42.4 in.^{3}. Find the radius of the jar to the nearest tenth of an inch.

1.5 in.

16. A playground in the shape of a right triangle has a length of 100 ft and an 80-ft side. Find the area of the playground.

4,000 ft^2

Use with text pages 483–485.

Name ______________________

Changing Areas

For Exercises 1–3, use a rectangle with a perimeter of 20 ft.

1. Complete the table.

Length (in ft)	0.5	1	1.5	2	2.5	3	3.5	4	4.5	5	5.5
Width (in ft)	9.5	9	8.5	8	7.5	7	6.5	6	5.5	5	4.5
Area (in ft^2)	4.75	9	12.75	16	18.75	21	22.75	24	24.75	25	24.75

2. Sketch a graph showing how the lengths and areas are related.

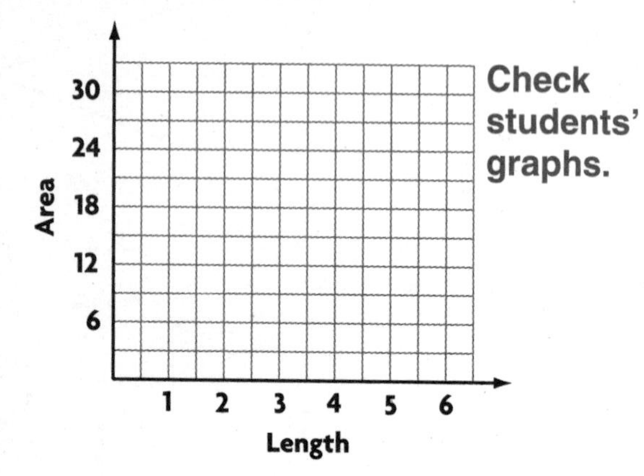

3. Sketch a graph showing how the lengths and widths are related.

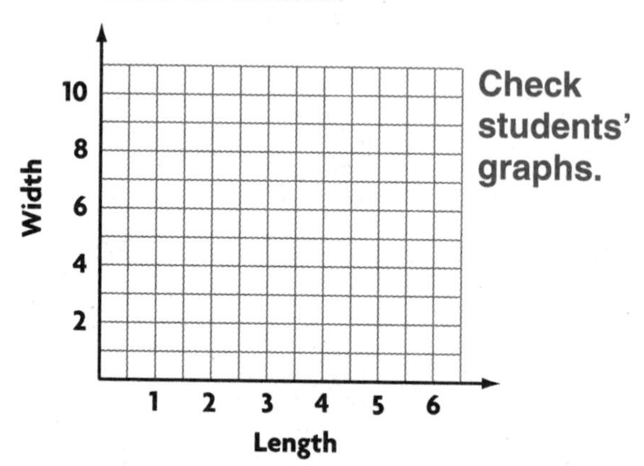

In Exercises 4–7, the perimeter of a rectangle is given. Using dimensions to the nearest 0.5 unit, find the length and width that will give the rectangle the largest possible area.

4. 60 mi 15 mi × 15 mi

5. 32 yd 8 yd × 8 yd

6. 150 in. 37.5 in. × 37.5 in.

7. 120 m 30 m × 30 m

Mixed Applications

8. A rectangular sandbox has an area of 36 ft^2. What are some possible perimeters for the sandbox?

Possible answers: 24 ft, 26 ft, 30 ft

9. Melanie has a garden in the shape of a circle, with a diameter of 6 ft. What is the area of her garden?

28.26 ft^2

10. A carpet maker has 148 yd of edging for a rug. What should be the length and width of the rug for the maximum rectangular area?

37 yd × 37 yd

11. A hot-air balloon is 34 m above the ground. A tree is directly under the balloon. Sandy is standing 16 m from the tree. How far are Sandy's feet from the balloon?

37.58 m

Use with text pages 491–493.

Name ____________________

Making Changes with Scaling

For Exercises 1–4, use the given scale selections to find the new length and width of each rectangle. Round to the nearest half inch.

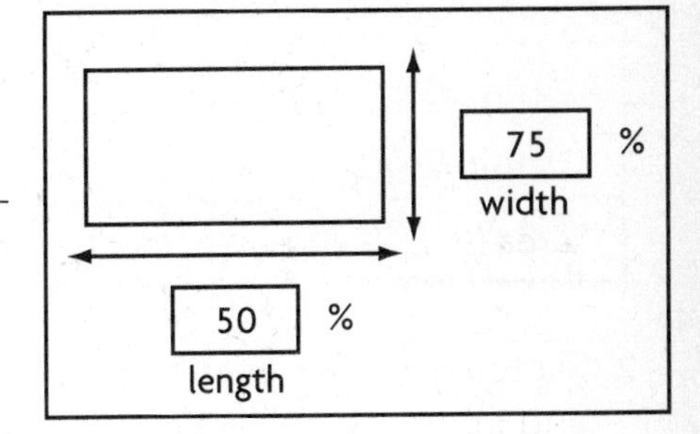

1. original length: 15 in.
   original width: 12 in.

   7.5 in.; 9 in.

2. original length: 32 in.
   original width: 44 in.

   16 in.; 33 in.

3. original length: 40 in.
   original width: 24 in.

   20 in.; 18 in.

4. original length: 25 in.
   original width: 30 in.

   12.5 in.; 22.5 in.

For Exercises 5–7, use the dimensions of the given rectangle and the scale selections.

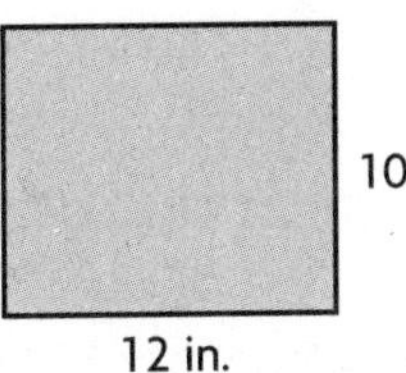

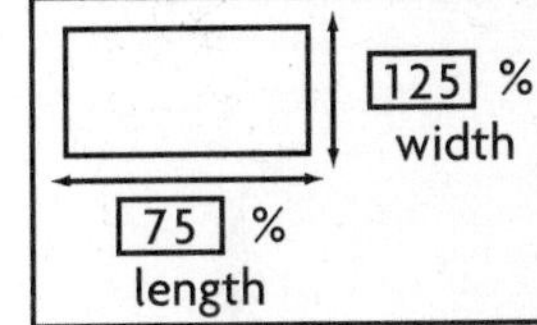

5. What are the new length and width?

   length 9 in.; width 12.5 in.

6. Find the perimeter and area of the new rectangle.

   perimeter 43 in.; area 112.5 in.^{2}

7. Find the increase or decrease from the original perimeter. **decrease of 1 in.**

For Exercises 8–9, use the given scale selections.

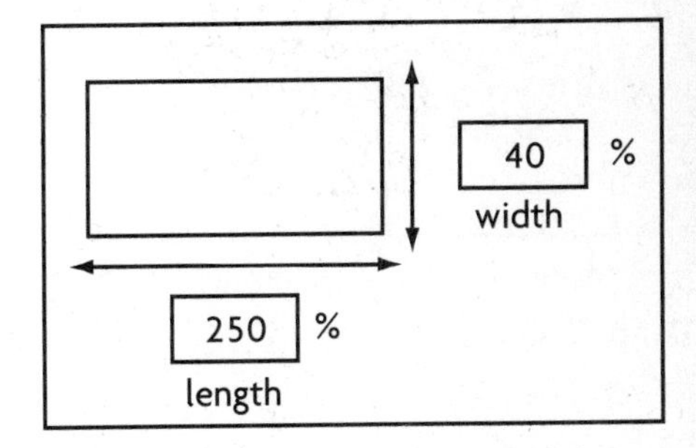

8. When this scale selection is applied to a rectangle 8 in. × 12 in., what happens to the area?

   The area does not change.

9. Give the length and width of a rectangle that does not show this relationship.

   There are no rectangles that do not show this relationship.

Mixed Applications

10. Max is shopping for computer diskettes. He can buy 12 diskettes for $27 or 15 diskettes for $33. Which is the better buy?

    15 diskettes for $33

11. A rectangular label is 5 in. long and 4 in. wide. The length is scaled to 120%. What scale setting for the width will make the perimeter of the new rectangle the same as the perimeter of the original rectangle?

    75%

Name ______________________

Problem-Solving Strategy

Making Models: Volume and Surface Area

Materials needed: graph paper

Make models to solve.

1. Crunch Pop cereal comes in two sizes. One box has a length of 8 in., a width of 2 in., and a height of 10 in. The other box has a length of 5 in., a width of 6 in., and a height of 5 in. Which box will hold the greater volume of cereal?

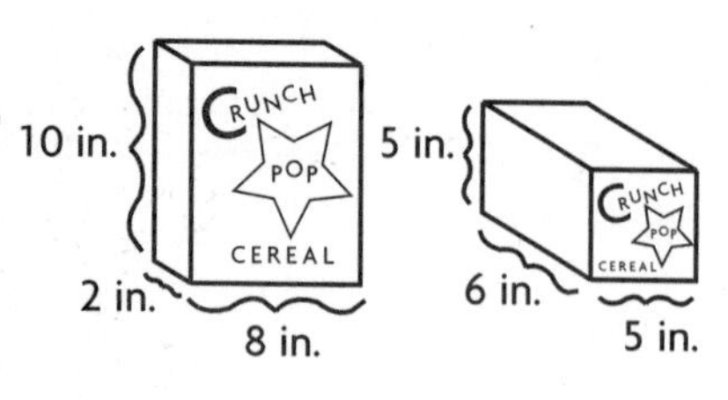

the box that is 8 in. × 2 in. × 10 in.

2. If the cost of making Crunch Pop boxes is $\frac{1}{2}$ cent per square inch, how much will each size cost to produce?

The 8 in. × 2 in. × 10 in. box will cost $1.16; the 5 in. × 6 in. × 5 in. box will cost $0.85.

3. What is the greatest number of Crunch Pop boxes you can display on a shelf that is 11 in. tall, 6 in. wide, and 36 in. long?

14 boxes that are 5 in. × 6 in. × 5 in.

4. A new lamp is packaged in a rectangular box 15 in. wide, 15 in. long, and 15 in. tall. How many of these boxes can fit in a crate 5 ft wide, 5 ft long, and 5 ft tall?

64 boxes

Mixed Applications

Solve.

CHOOSE A STRATEGY

- Make a Model
- Draw a Picture
- Make a Table
- Use a Formula
- Write an Equation
- Guess and Check

Choices of strategies will vary.

5. Health-Crisp potato chips come in a cylindrical can with a diameter of 3 in. and a height of 8 in. How many of these cylinders can fit in a box 2 ft long, 2 ft wide, and 2 ft high?

128 cylinders

6. Larry's Light Company requires 95% of the night lights it produces to light on the first click of the switch. If 220 lights are randomly tested and 205 light on the first click, has the company met its requirement? Explain.

no; $0.95 \times 220 = 209$; $205 < 209$

7. The difference of two numbers is 10. When added together the sum is 52. What are the two numbers? Find the product of these two numbers.

21, 31; 651

8. Sarah wants to save $256 in 8 months. After 3 months, she has saved $105. At this rate, will Sarah reach her goal? How far above or below her goal will she be?

yes; $24 above her goal

Use with text pages 498–500.

Name ______________________

Volumes of Changing Cylinders

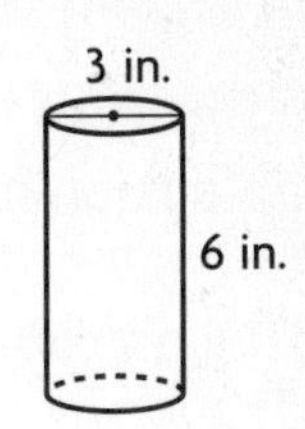

For Exercises 1–6, use the model of a cylinder shown here.

For Exercises 1–3, the given scale factor was used to create the model. Find the radius, height, and volume of the original cylinder. HINT: To find the radius, use this equation: scale factor × original radius = model radius.

1. $\frac{1}{3}$ **4.5 in.; 18 in.; 1,145 in.^{3}**

2. $\frac{2}{3}$ **2.25 in.; 9 in.; 143 in.^{3}**

3. $\frac{5}{4}$ **1.2 in.; 4.8 in.; 22 in.^{3}**

4. In which exercise(s) did the scale factor decrease the volume of the original cylinder? **1 and 2**

5. In which exercise(s) did the scale factor increase the volume of the original cylinder? **3**

6. Use a calculator to compare the volume of the model with the volume of the original cylinder in Exercise 5. **The ratio is approximately equal to the scale factor cubed.**

Using the cylinders below, complete the table. Round volumes to the nearest cubic foot.

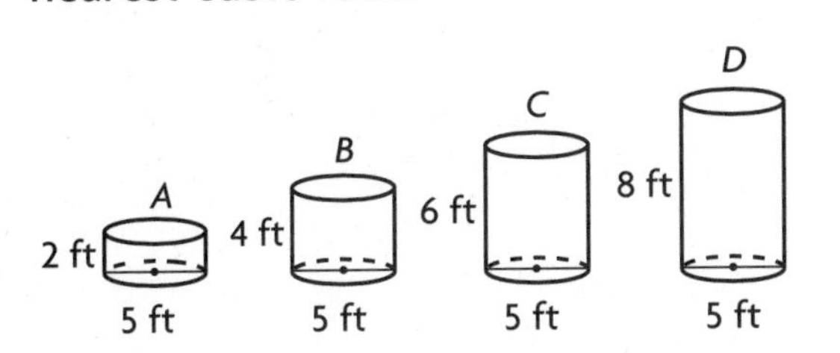

		Cylinder *A*	Cylinder *B*	Cylinder *C*	Cylinder *D*
7.	Radius (in ft)	**2.5 ft**	**2.5 ft**	**2.5 ft**	**2.5 ft**
8.	Height (in ft)	**2 ft**	**4 ft**	**6 ft**	**8 ft**
9.	Volume (in ft^3)	**39 ft^3**	**79 ft^3**	**118 ft^3**	**157 ft^3**

10. For these cylinders, sketch a graph relating the height and the volume. **Check students' graphs.**

11. Based on your graph and the table, what are your conclusions?

    Volume increases as height increases.

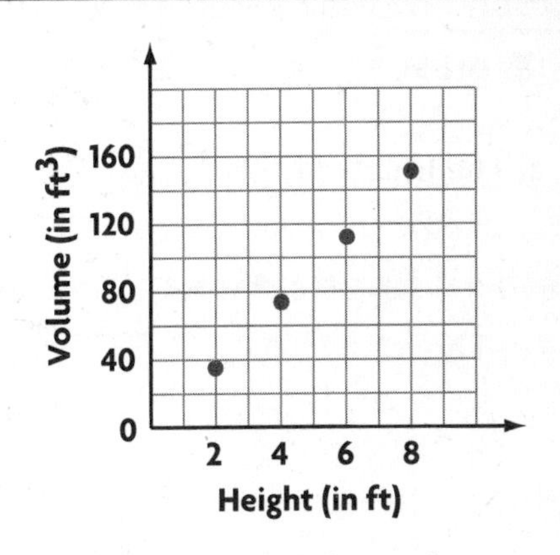

Mixed Applications

12. For the above cylinders, find the volumes of a model of cylinder *B* using a scale factor of $\frac{7}{5}$ and a model of cylinder *A* using a scale factor of $\frac{8}{5}$. Which model has the greater volume?

    ≈ 158 ft^3; ≈ 161 ft^3; cylinder *A*

13. The Bates family is trying to keep deer out of the garden. They buy 96 ft of fencing. What is the greatest area that can be enclosed by the fence?

    576 ft^2

Name ______________________

LESSON 27.1

Percent and Sales Tax

Find the sales tax and the total cost. Describe how you can do the computation mentally.

1. 8% on $50.00 $4; $54; multiply price by tax rate, then add to price.

2. 4% on $900.00 $36; $936; multiply price by tax rate, then add to price.

Find the total cost directly. Round to the nearest cent when necessary.

3. 3% on $35.76 $36.83

4. 7.5% on $150 $161.25

5. 9% on $56.29 $61.36

6. 7% on $78.60 $84.10

Find the price. Round to the nearest cent when necessary.

7. 5% sales tax

$10.92 total cost $10.40

8. $7\frac{3}{4}$% sales tax

$70.47 total cost $65.40

9. $4\frac{1}{2}$% sales tax

$78.38 total cost $75

Tell the number you would divide by to find the price directly from the total cost.

10. 7% sales tax 1.07

11. 5.5% sales tax 1.055

12. $6\frac{1}{2}$% sales tax 1.065

Find the sales tax rate for the given price and amount of sales tax. Round to the nearest tenth of a percent.

13. price: $36.00

sales tax: $1.08 3%

14. price: $75.99

sales tax: $4.94 6.5%

15. price: $6.75

sales tax: $0.54 8%

Mixed Applications

16. The total cost of a winter jacket, including sales tax, is $57.20. Find the price if the sales tax rate is 7%.

$53.46

17. The price of a pair of shoes is $45.50. Find the total cost if the sales tax rate is 6.5%.

$48.46

18. A building lot with an area of 21,875 ft^2 is in the shape of a right triangle. If one leg of the triangle measures 350 ft, find the length of the other leg.

125 ft

19. The mean of three numbers is 213. Two of the numbers are 52 and 85. Find the third number.

502

Use with text pages 512–515.

Name ______________________

Percent and Discount

Vocabulary

Complete.

1. The amount an item is marked down and the amount the customer will save is the **discount**.

2. The new price after an item is marked down is the **sale price**.

Find the sale price.

3. Regular price: $30

S•A•L•E 20% OFF **$24**

4. Regular price: $48.00

25% Discount **$36**

5. Regular price: $78.50

SAVE 10% **$70.65**

Find the regular price.

6. Sale price: $2.25

• 25% off **$3**

7. Sale price: $46.80

35% off **$72**

8. Sale price: $135

40% DISCOUNT **$225**

Find the rate of discount.

9. Regular price: $27.00

Sale price: $20.25 **25%**

10. Regular price: $5.40

Sale price: $4.32 **20%**

11. Regular price: $68.00

Sale price: $44.20 **35%**

Find the total cost of each item. The sales tax rate is 6%.

12. Regular price: $19.50

Now 20% off **$16.54**

13. Regular price: $76.00

Now 25% off **$60.42**

14. Regular price: $6.50

Now 10% off **$6.20**

Mixed Applications

15. Griffen buys a $475.00 computer on sale for 20% off. If the sales tax rate is 6.5%, what is the total cost of the computer?

$404.70

16. John is 3 years older than Jerry. The sum of their ages is 21. How old is Jerry? How old is John?

9 yr; 12 yr

17. Lou Ellen buys a computer game and CD that regularly cost $29.90 and $15.99. She pays a total of $40.96, which includes a sales tax of $1.95. What are the sales tax rate, discount, and rate of discount?

5%; 15%, $6.88

18. Oren is painting the walls of his room, which have an area of 720 ft^2. One quart of paint covers 160 ft^2. A gallon of paint costs $12.00, and a quart costs $3.75. How much does Oren pay for paint?

$15.75

Name ____________________

Percent and Markup

Vocabulary

Write the correct letter from column 2.

1. wholesale price __b__
2. retail price __d__
3. markup __a__
4. profit __c__

a. difference between wholesale price and retail price
b. price at which a store buys merchandise in large quantities
c. money made over cost on items sold
d. price at which a store sells merchandise

Find the value of the markup.

5. wholesale price: $45.20
   retail price: $89.95 __$44.75__
6. wholesale price: $8.50
   retail price: $11.85 __$3.35__
7. wholesale price: $60.00
   retail price: $99.99 __$39.99__

You buy computers at a wholesale price of $450 each. Find the value of your markup and retail price if you sell the computers with these markups.

8. 100% markup __$450; $900__
9. 75% markup __$337.50; $787.50__
10. 140% markup __$630; $1,080__
11. 110% markup __$495; $945__

Find the amount and percent of each markup.

12. wholesale price: $40
    retail price: $100 __$60; 150%__
13. wholesale price: $8
    retail price: $12 __$4; 50%__
14. wholesale price: $35
    retail price: $70 __$35; 100%__

Mixed Applications

15. A store's wholesale prices are $9.50 for sweatshirts, $2.40 for socks, and $18.00 for sweatpants. The markup is 75%. Find the retail prices.

    __sweatshirts: $16.63;__
    __socks: $4.20; sweatpants: $31.50__

16. A store's markup is 125% of the wholesale price. The retail prices are $16.88 for a box of disks, $22.50 for a telephone, and $3.75 for a calendar. What was the wholesale price of each item?

    __disks: $13.50; telephone: $18;__
    __calendar: $3__

17. If a car travels at an average speed of 45 mi per hour, how long will it take to complete a trip of 270 mi?

    __6 hr__

18. Andrea ran 2 miles in 17 minutes. At that rate, what was her time for each half mile?

    __4.25 min__

Use with text pages 520–523.

Name ___________________________

Earning Simple Interest

Vocabulary

Write the correct letter from column 2.

1. principal <u>b</u>
2. total amount <u>c</u>
3. simple interest <u>a</u>

a. amount you earn computed only on the amount you deposit

b. amount you deposit

c. interest plus principal, or A in the equation $A = p + I$

For Exercises 4–9, assume $800 is deposited at 5% simple interest. Complete the tables to show growth over a five-year period.

	Year	Interest	Amount
	Start		$800
	1	$40	$840
4.	2	$40	$880

	Year	Interest	Amount
5.	3	$40	$920
6.	4	$40	$960
7.	5	$40	$1,000

8. When the $800 is deposited for t years, you can compute the interest earned with the expression $800(0.05)$t$. Write an expression for the total amount at the end of t years.

   $800 + $800(0.05)$t$

9. Use the expression from Exercise 8 to find the interest and total amount when $t = 4\frac{1}{2}$ years.

   $180; $980

Use your calculator to find the interest.

10. $4\frac{1}{2}$% simple interest for 3 years on a principal of $550.00

    $74.25

11. 4% simple interest for 2 years on a principal of $176.00

    $14.08

12. 8.25% simple interest for 10 years on a principal of $5,600.00

    $4,620

Mixed Applications

13. Jefferson deposits $900.00 in an account that earns 4% simple interest. In how many years will the total amount equal $972.00?

    2 yr

14. Kevin earned $156.25. If his hourly pay is $6.25 per hour, how many hours did he work? How many hours will he have to work to earn $200?

    25 hr; 32 hr

Name ______________________

LESSON 27.5

Problem-Solving Strategy

Making a Table to Find Interest

Vocabulary

Complete.

1. If you do not pay the full amount on a credit card bill, interest called a(n) finance charge will be added to the new balance.

Make a table and solve.

2. Mrs. Meier charged $65.40 on her credit card for CDs. The interest rate is 12% annually, or 1% monthly. She will make a $20.00 payment each month. How much interest will she pay? What is the total amount she will pay?

Month	Interest	Balance	Payment	New Balance
1	$0.00	$65.40	$20.00	$45.40
2	$0.46	$45.86	$20.00	$25.86
3	$0.26	$26.12	$20.00	$6.12
4	$0.07	$6.19	$6.19	$0.00
Total	$0.79		$66.19	

$0.79; $66.19

Mixed Applications

Solve.

CHOOSE A STRATEGY

- Make a Table
- Write an Equation
- Use a Formula
- Guess and Check
- Make a List
- Work Backward

Choices of strategies will vary.

3. Lucien charged $102.40 on his credit card. The interest rate is 18% annually. He paid $27.40 the first month and $25.00 each month after that. What is the monthly interest rate? What is the total amount he will pay?

$1.5%; $104.73

4. A chess tournament includes 7 players. There are 4 players over 12 years of age and 3 players under 12. Every player plays every other player once. How many games are played between a player over 12 and a player under 12?

12 games

5. Richard charged $55.20 on his credit card for clothing. The interest rate is 15% annually, or 1.25% monthly. If he pays $20.00 a month, how long will it take him to pay the amount charged? How much interest will he pay? What is the total amount he will pay?

3 mo; $0.64; $55.84

6. Rex is building a circular table with a diameter of 28 in. What is the circumference? What is the area of the table top? Use $\pi = 3.14$. Round to the nearest inch.

88 in.; 615 in.^{2}

Use with text pages 527–529.

Name ______________________

LESSON 28.1

Graphs and Pictures

The picture at the right shows the path of a person jumping over a hurdle. Use this picture for Exercises 1–2.

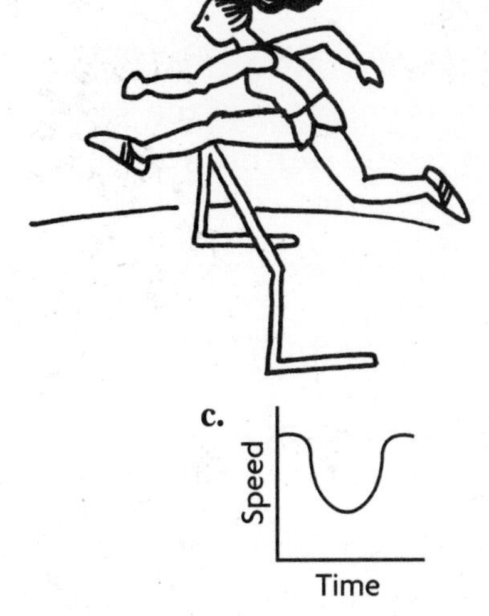

1. Which graph below shows the relationship between time and the speed of the person jumping over the hurdle? __c__

a. 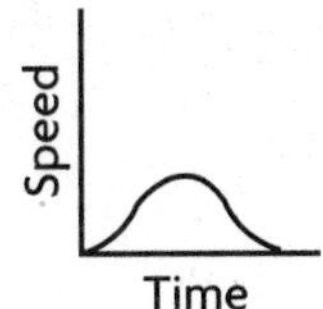

b. 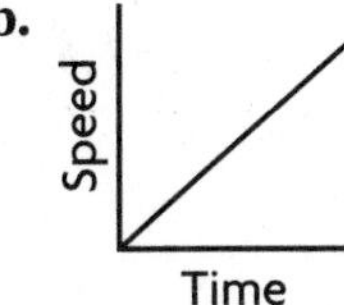

c. 

2. Draw a graph that shows the relationship between time and the distance the hurdle jumper travels. **Check students' drawings. Graph should show that distance increases as the time increases.**

3. Which graph below illustrates the relationship between time and the distance between a ball that has been pitched and a batter shown in the picture at the right? Explain.

__a; the graph shows that as time increases, the dist__

__between the ball and the batter decreases.__

a. 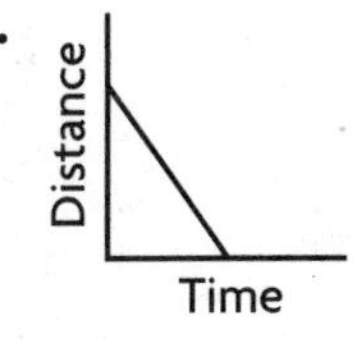

b. 

c. 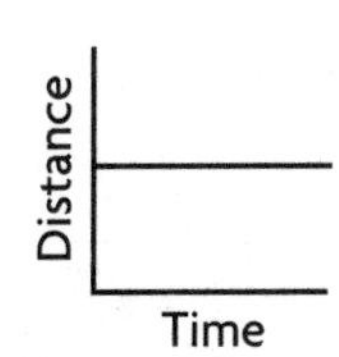

4. Which graph for Exercise 3 illustrates the relationship between time and the distance between a merry-go-round horse and the center of the merry-go-round? Explain.

__c; the graph shows that as time increases, the distance__

__between a horse and the center of the merry-go-round__

__stays the same.__

Mixed Applications

5. Draw a picture of an event where speed increases as time increases.
Check students' pictures.

6. In how many ways can Coach Jones choose a team of 3 wrestlers from a group of 5 wrestlers?

__10 ways__

Name ________________________________

Relationships in Graphs

1. Look at the graph at the right. Describe how the air temperature changed as the time increased in the 24-hr period starting from midnight.

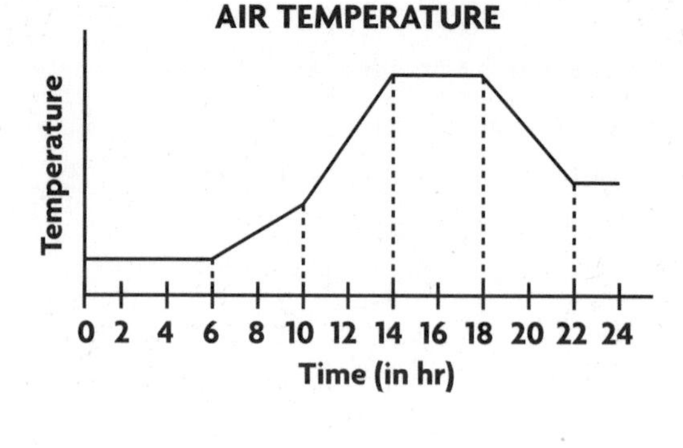

The temperature stayed the same,

increased, increased more,

stayed the same, decreased,

and stayed the same.

2. The graph at the right shows the relationship between the temperature inside an oven and the time needed to bake a cake. Tell a story about baking the cake. **Check students' stories.**

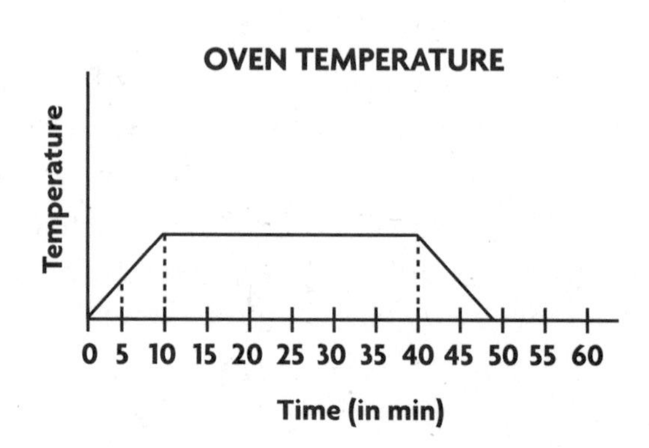

3. Which part of the graph represents the oven's heating up?

   the increasing part

4. Which part of the graph represents the oven's cooling down?

   the decreasing part

5. Which part of the graph represents baking the cake?

   the constant interval from 10–40 min

Mixed Applications

6. The graphs at the right show relationships in a bicycle race. For which interval do both the distance and the speed stay constant?

   4–6 min

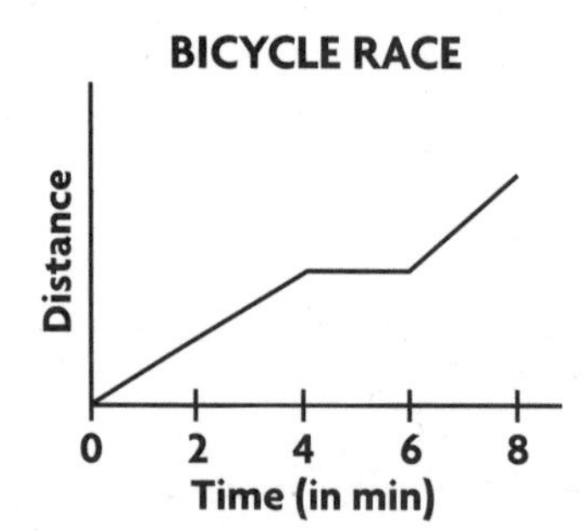

BICYCLE RACE

Speed

0 2 4 6 8

Time (in min)

7. A rectangular park is 500 m long and 375 m wide. Katie walked diagonally across the park from corner to corner. How far did she walk?

   625 m

8. A survey shows that 75% of the seventh-grade students at Marker Junior High like pizza. If 186 students like pizza, how many students were surveyed?

   248 students

Use with text pages 536–539.

Name ______________________________

Graphing Relationships

Identify the variables you would use to graph the relationship.

1. The greater the number of years employed, the higher the salary.

   number of years employed, salary

2. The larger the bottles, the fewer the bottles that fit in a carton.

   size of bottle, bottles per carton

3. The greater the number of ounces a gold necklace weighs, the more it costs.

   weight, cost

4. The greater the patient's age, the greater the dose of cough syrup.

   age, size of dose

Complete the graph.

5. The pressure in a swimmer's ears increases as the underwater depth increases.

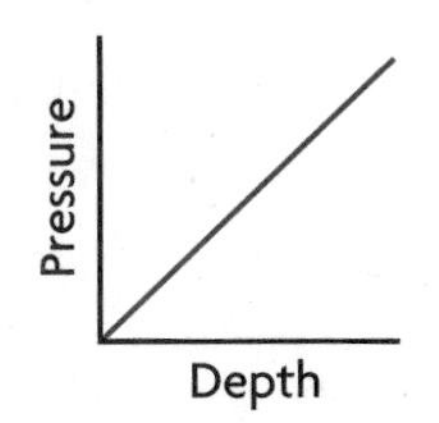

6. The greater the age of an automobile, the lower the resale value.

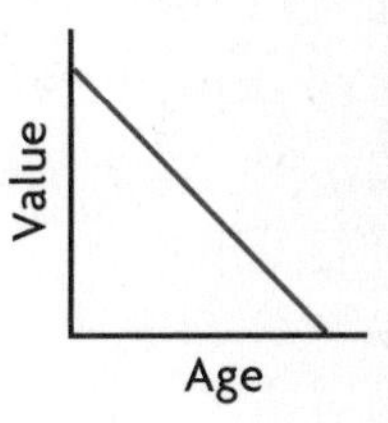

Sketch a graph of the relationship.

7. The amount of the discount increases as the amount you spend increases.

8. The farther you are from a light source, the dimmer the illumination.

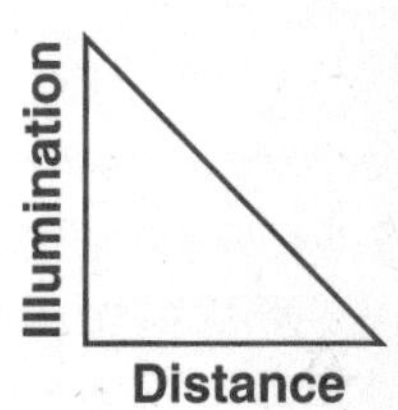

Mixed Applications

9. Write an example of a relationship where one variable stays constant as the other variable increases. Then draw a graph for the relationship.

   Check students' examples and graphs.

10. Jackie wants to phone her pen pal in Japan. The call costs $4.50 for the first 3 min and $0.95 for each additional minute or part of a minute. How long can Jackie talk if she has $8.65?

    7 min

11. Jhvonne cuts a 60-ft rope into four pieces. Each successive piece is twice as long as the previous piece. How long is the longest piece?

    32 ft

Name ______________________

LESSON 28.4

Using Scatterplots

Vocabulary

Write the letter of the correct description from Column 2.

1. scatterplot ___c___
2. positive correlation ___d___
3. negative correlation ___b___
4. no correlation ___a___

a. no pattern can be formed
b. pattern slants downward in a straight line
c. a graph of ordered pairs
d. pattern slants upward in a straight line

Write *positive, negative,* or *no correlation* to describe the relationship between the two sets of data.

5. length of a phone call; cost of the phone call

positive

6. person's hair color; person's shoe size

no correlation

7. distance traveled to school; distance left to travel

negative

8. scoring average; number of games won

positive

Draw a scatterplot to represent the set of data. Write *positive, negative,* or *no correlation* to describe the relationship.

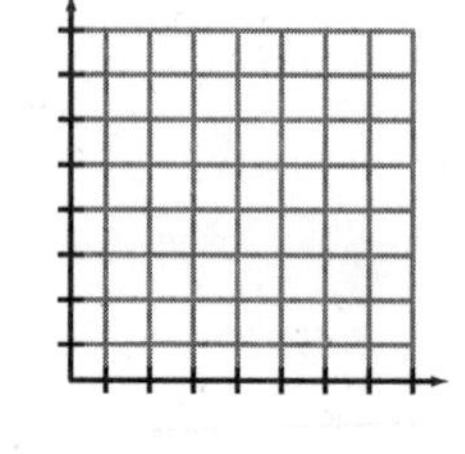

9.

Time (in min)	1	2	3	4	5	6	7
Depth (in cm)	1	1	2	3	3	4	6

Check students' graphs.

positive

Mixed Applications

10. In the scatterplot at the right, what is the relationship between the number of hours spent studying and the test scores?

As the number of study hours increased, the test scores increased.

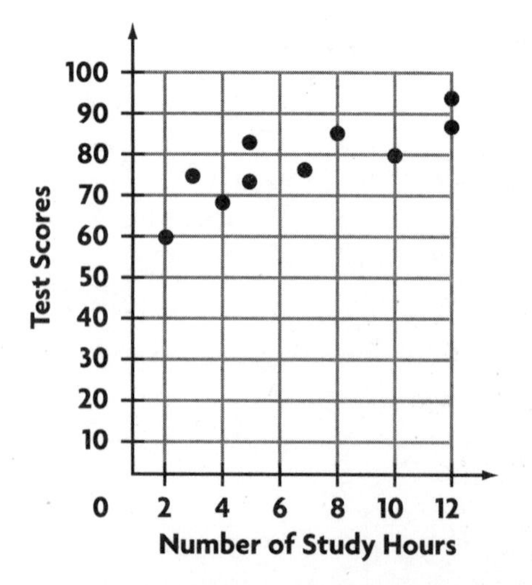

11. A TV and a VCR together sell for \$655. If the TV costs \$55 more than the VCR, how much does the TV cost?

\$355

Use with text pages 543–545.